be returned on or before
date stamped below.

COMPUTER NUMERICAL CONTROL

Joseph Pusztai
and
Michael Sava

Reston Publishing Company, Inc.
A Prentice-Hall Company
Reston, Virginia

Library of Congress Cataloging in Publication Data

Sava, Michael
      Computer numerical control.

      1. Machine-tools-—Numerical control.   I. Pusztai,
Joseph.   II. Title.
TJ1189.S28   1983   621.9′023   82-25039
ISBN 0-8359-0924-7

© 1983 by Reston Publishing Company, Inc.
A Prentice-Hall Company
Reston, Virginia 22090

10 9 8 7 6 5 4 3 2 1

Printed in the United States of America

Interior Designed by Dan McCauley

Contents

Chapter 4　　Machining Forces, 57

Chapter 5　　Cutter Centerline Programming, 67

Preface

This book is the result of our continually increasing interest in numerical control over nearly two decades. It reflects the summary of notes prepared for industrial seminars, operators and programmers, tool-and-die makers and machinists, technicians and technologists, as well as for night school courses; it also contains detailed programs for some interesting parts produced for industrial clients.

The contents have been successfully tested as a self learning package and as conventional lecture material.

The book focuses on graphical presentation, and it consists essentially of a large number of worked out examples, together with fundamental rules and general principles, derived essentially from experience.

The text is organized in self-contained chapters, containing illustrations, explanations and comments. It should fulfill specific needs of people studying for careers in the computer numerical control (CNC) field, whether in technical institutes and colleges, or as self-learners in industry. It should also allow instructors to follow their particular order of presentation, to suit specific terminal objectives.

The book is also designed to help those now working in planning, tool design, CNC maintenance, or as junior programmers and operators.

The first four chapters represent a gradual introduction to the mathematics and codes of numerical control, to the principles of the system and to the forces resulting from the tool-part interaction.

Chapter 5 is a thorough review of conventional numerical control, which introduces the beginner to the basics of conventional non-CNC programming. The reader has noticed at this stage the interchangeable use of computer numerical control (CNC) and numerical control (NC). The latter means the older, non-computerized, hardwired version, and the former is the all-encompassing microprocessor controlled, softwired system of the present. The two terms are used interchangeably in industry and should not generate any confusion. Chapter 5 will serve the needs of older equipment and it will ease the way into CNC by introducing some of the objectives, principles and vocabulary.

Any new technology suffers from a lack of terminology standardization. Chapter 6 is an attempt in this direction. It deals with tool offsets and their applications, as distinct from tool diameter or length compensation which are discussed extensively in chapters 7 and 8.

Chapter 9 discusses the issue of canned cycles, supplied by the manufacturer as either standard or optional equipment. It also introduces a newer feature—the user designed canned cycle. The latter is illustrated extensively in chapter 11, following the description of some additional CNC features in chapter 10. The preceding 11 chapters of *Computer Numerical Control* have nevertheless dealt with what people in industry call *manual* part programming.

Chapter 12 concludes this book with a comprehensive introduction to two major languages of *computer*-assisted part programming, as understood in the field.

No single book will ever fill all the needs of those who study it. It is hoped that the readers will be able to transfer the knowledge acquired to the systems of their choice without any great difficulty.

We have been fortunate to become members of an educational organization among the most progressive in the field of technical training and one of the best equipped in CNC. We would like to thank the administration of Humber College for its decisive approach to high technology.

We are grateful to our students from both college and industry for their time, efforts and challenging questions on the materials that eventually led to this book. Recognition should also be extended to the efforts of the full-time and part-time NC teaching staff, Larry Barnard, who tested many of the programs shown; Alex Goldie and Frank Franklin; John Wallace of MDSI; Steve Pereira of CNC Programming Services; as well as Helen Stojanovic, who prepared illustrations for many of the courses this book was based on.

We would also like to express our appreciation to our colleagues and friends, too numerous to mention here, from both the manufacturing industry and the machine-tool distributors, for information supplied, for their help, cooperation and many useful suggestions. Lastly, we would like to thank Livia Pusztai and Rena Sava for their patience, understanding and encouragement.

Joseph Pusztai
Michael Sava

The Development
of Numerical
Control

The evolution of the machine tool industry could hardly be appreciated without a brief review of its birth and growth. John Wilkinson built his metal-cutting boring machine in the eighteenth century, but nearly two centuries of evolution were needed to produce the hydraulic tracer-controlled copy mills and lathes. The next stage, automation, was brought about by mass production of automobiles, agricultural implements, household appliances, chemical products, as well as inventory and financial data handling. Three kinds of automation met the needs of society for a major part of the twentieth century. These were:

1. Automotive or fixed assembly line automation (Detroit type)
2. Process control automation, primarily used in the manufacture of chemical and food products
3. Data processing, first developed for processing payrolls, data collection, and inventory control.

The Second World War marked the turning point in the ability of the metal-cutting industry to cope with the requirements facing it. The ambitious aircraft and missile projects of the U.S. airforce, combined with the demands for commercial jets, made it quite clear that conventional manufacturing could not fulfill future needs. A study of the U.S. government showed that the combined resources of the entire U.S. metal-cutting industry in 1947 could not produce the parts needed by the airforce alone.

Under contract to the U.S. airforce, the Parsons Corporation undertook the development of a flexible, dynamic manufacturing system, designed to maximize productivity by emphasizing details required to achieve desired accuracies. This system would allow design changes without costly modifications to tooling and fixturing, and it would fit into a modern, productive manufacturing management for small-to-medium sized production runs. The Parsons Corporation subcontracted the development of the control system to the Massachusetts Institute of Technology (MIT) in 1951. A control, which could be applicable to a wide variety of machine tools, would drive a slide lead screw through an interface, as instructed by the output of a computer. MIT met the challenge successfully, and in 1952 demonstrated a Cincinnati Hydrotel milling machine equipped with the new technology, which was named Numerical Control (NC) and used a prepunched tape as the input media. Since 1952, practically every machine tool manufacturer in the Western world has converted part or all of its product to NC.

The first NC machines used vacuum tubes, electrical relays, and complicated machine-control interfaces. The second generation of machines utilized improved miniature electronic tubes, and later solid-state circuits. As computer technology improved, NC underwent one of the most rapid changes known in history. The third generation used much improved integrated circuits. Computer hardware became progressively less expensive and more reliable and NC control builders introduced for the first time Read Only Memory (ROM) technology. ROM was typically used for program storage in special-purpose applications, leading to the appearance of the computer numerical control (CNC) system. CNC was successfully introduced to practically every manufacturing process. Drilling, milling, and turning were performed on "machining centers" and "turning centers." CNC took over glass cutting, pattern making, electrical discharge machining, steel-mill roll grinding, coordinate measuring, electron beam welding, tube bending, drafting, printed circuit manufacturing, coil winding, functional testing, robots, and many other processes.

A set of preprogrammed subroutines, named canned cycles, were developed for use in routine operations. They were recorded into the ROMs and remained there even after power was shut off. For the first time, this concept made it possible to read the machining program into memory and to operate the machine from memory. In addition to the advantages of editing, the problems caused by erroneous tape reading disappeared.

Along with the many canned cycle options, CNC builders introduced displays for visual editing of part programs in memory. Various in-cycle problems generated alarms and hundreds of diagnostic messages which could be displayed as applicable. Practically every function of the machine was tied into the system and monitored during operation. A constant surface speed control was incorporated and continuously anticipated the most efficient spindle speed for the next cut to minimize time lost for spindle acceleration. The conventional linear and circular interpolation in cartesian (rectangular) coordinates were supplemented by polar coordinates and helical interpolation. Safe zones, which could be set through programmed codes or internal parameters, created an electronic crash barrier to prevent tool collision. The latter group of features marked the arrival of high technology to the manufacturing or metal-cutting industry.

The improvement in drives was as important for the system as the contribution of the microprocessor or the minicomputer.

The feed drives, usually known as servodrives, consist of a motor and its control which receives its motion instructions from the CNC. Their performance is essential to the accuracy, reliability, and flexibility of the CNC system.

The open-loop system is normally used in simple point-to-point, or positioning systems, although improvements in technology have made it possible to install the system in contouring systems as well.

The closed-loop configuration is more accurate and reliable, as reflected by its higher cost.

Although many CNC systems still use hydraulic or pulse motors, the DC drives have gained dominance on a much larger scale. In most cases, the drive packages are purchased from specialized drive system builders. These direct current (DC) permanent-magnet wound field servomotors range from 3,000 revolutions per minute (rpm) to less than 1 rpm without stalling. They develop peak torque capabilities with high slide acceleration and low inertia for optimized system response. Most drive systems offer a choice between transistorized silicone-controlled rectifiers and pulse width modulation over the full range of amplified voltages. These drives can now drive virtually any lead screw. Their high-response inner current loops provide reliable regulation of torque-load disturbances. They can also be built with high-gain preamplifiers to close high-bandwidth velocity loops. The DC drives provide the answer to the most essential needs of acceleration, deceleration, stopping, and constant velocity, with inherent shaft stiffness for successful operation of the CNC system. The same drive systems actuate robots, transfer lines, flight simulators, graphic plotters, etc. As these drives are infinitely variable and fully regenerative, they can provide for maximum performance and control over the whole range of the motor. By eliminating gearboxes and clutches, the cost of drives for the third-generation CNC systems was reduced substantially.

The fourth-generation microprocessor CNC incorporated in many cases the controversial bubble memory. The bubbles are magnetic garnet crystals grown on nonmagnetic substrate, ranging in size from 2 to 30 micrometres, and used as non-volatile data storage. Although at this stage it is not competitive in the large computers, the bubble memory is closing the cost gap with disk storage devices. Insensitive to adverse temperature changes, dust, and vibration, the bubble memory has demonstrated superior reliability in shop environment. General Numerics introduced its fourth-generation CNC using bubble memory; however Hitachi, another electronic giant, believes that bubble memory will provide the economical answer to direct numerical control (DNC).

Among the strengths of the fourth-generation microprocessor CNC (MCNC) are added part program memory storage, reduction of printed circuit boards, programmable interface, faster memory access, parametric subroutines, and macro capabilities.

The system user can now write specific canned cycles directed to particular applications ("user macros"), far more economical and efficient than conventional canned cycles. Mathematical calculations with do-loop subroutines using variables can now be incorporated in the part program. The microprocessor controls both computations and motion commands. Thus, following an in-process gaging, an out-of-tolerance condition will be fed back, and the tool offset will be automatically modified to achieve the desired part dimensions.

In addition, the fourth-generation microcomputer CNC system has the ability to control typical robot functions such as loading and unloading parts. Using the teach-in learning mode, the robot can be programmed to change tools or to remove chips.

The maintenance aspect, however, has not kept up with the system development. Many CNCs and MCNCs are not sufficiently protected against power surges, spikes, or transients, caused by welding equipment, solenoids, motors, or fluorescent lights connected to the same power supply as the CNC. Potential problems are slide runaways and loss or damage to programs in memory.

Many of the maintenance headaches could and probably will be eliminated by phasing out the fast-response, low-energy (metal oxide version) suppressors or the medium-response, high-energy (isolation transformers) suppressors. The replacement, solid-state suppressors, have a response time on the order of 4-6 nanoseconds, ten times faster than the metal oxide versions (MOVs). Their energy-dissipating capacity of 16,667 J is far higher than the 40-J capacity of the MOVs.

Where will technology go from here? To a large extent it depends on the knowledge of the system users and the demands they will pose to the designers and builders of manufacturing systems. CNC will probably remain for a long time one of the most practical elements of computer-aided design and computer-aided manufacturing technology.

2

Mathematics for Computer Numerical Control

The previous chapter emphasized the extraordinary capabilities of the CNC machine. However smart a CNC may be, it simply cannot think. It can perform unlimited numbers of activities and can repeat any number of operations continuously and consistently by the use of numeric directions and commands. In this chapter we will discuss two different types of mathematics: One which deals with and leads up to the tape codes, or Binary Coded Decimal (BCD) mathematics, and the other which is used by the programmer as an aid to calculate tool centerline dimensions. The latter type of mathematics, known as trigonometry and analytical geometry, will present useful formulas essential to every CNC programmer for writing efficient part programs.

2.1 THE NUMERICS THAT CNC MACHINE TOOLS UNDERSTAND

The basic hardware of the CNC consists of the input units, the computing or mathematics unit, the memory unit, the control unit, and the output units. The function of any input unit is to provide data to the computer in the form of numeric instructions. Present CNC systems are designed to operate with different input media. The most common of these is the punched tape, mainly because it can be read inexpensively, is less sensitive to handling, is inexpensive to purchase, and requires less equipment to make and less costly space for data storage. Its disadvantage, however, is that it cannot be reused. The magnetic tape has limited use as a CNC media, it requires sophisticated (expensive) equipment for program recording and reading, and the programmer or operator cannot see the recorded codes and therefore cannot read them. Recording errors are not as obvious and visible as they are in punched tape. Magnetic Tape requires special storage space and must be handled carefully to avoid erasing the program. The typewriter, more commonly known as the keyboard, has limited use because of the operator's speed. It may not be used for long program input, but is primarily for small programs. Its main use is to edit (correct) programs already in memory or to generate single operations in the Manual Data Input (MDI) mode. The objective here is to introduce the reader to the evolution of the codes we use in our input units to communicate with the CNC computer.

The part program, once read into the computer memory, becomes a set of instructions to perform calculations. The most popular CNC memories are still the Semiconductor IC memory and the Magnetic Core memory. However, "Bubble" memories are being used now in some CNC systems. The internal CNC memory can only handle small amounts of data, but at a very fast rate. It is not uncommon for computers to operate at 1 million words per second. Because of this rate of speed, CNC can perform linear, circular, and parabolic interpolation (calculations) at a rate of 200 to 300 inches per minute (ipm) slide velocities.

The "brain" or control unit of the CNC controls how these operations are performed. It translates the memory instructions and specifies what operations are to be performed in what sequence. The mathematical unit performs simple addition, subtraction, multiplication, and division functions. The results are fed back to the memory for storage, or read out to the various output units of the CNC. These output units are servodrives for slides or program readouts to teletype printer, CRT, or tape punch unit. Tool changes or other miscellaneous codes are not handled in the mathematics. The codes or numbering system of these calculations are different from the input code used in the punched tape. Later in this chapter we will discuss, in detail, how the different numbering systems function in the CNC.

The CNC is a special-purpose computer, using special CNC commands in a simplified manner called programming. These commands are written instructions in schematic form. The programmer does not have to describe in every detail what steps the CNC is to perform. A few commands such as we use to instruct the machine to cut a 360° circular arc will cause the computer to perform thousands of calculations involving additions, subtractions, multiplications, and divisions. A CNC that cannot do these types of calculations would have very little use in complex parts manufacturing.

The accuracy of the calculations is limited by the digits used for each number. In most CNCs the number length, called the number of binary digits, is fixed to 8 and 16 bits (binary digits). The precision of the CNC far exceeds the physical limitations of the mechanical devices such as lead screws and slides. In spite of all the glorious things said about the CNC, it is the programmer who does the thinking and achieves the precision. The CNC is a very primitive piece of hardware; it can only understand numbers composed of "1" and "0," in electrical terms "on" or "off," sensing the presence or absence of magnetism or voltage.

2.1.1 Numbering Systems: Decimal Number System

In our every-day life we seldom think of, or analyze, the numbers we use and work with. For example, the number 649 should really be written as $649_{(10)}$ meaning base 10. The base 10 system has digits from 0, 1, 2, 3, 4, . . . 9 but there is no 10. The 10 is not a basic digit in the system. The second important point to note is that the position of the digit in the number defines the value of units, tens, hundreds, thousands, etc.

The number

$$649_{(10)} = 9 \times 10^0 + 4 \times 10^1 + 6 \times 10^2$$
$$= 9 \times 1 + 4 \times 10 + 6 \times 100$$
$$= 649_{(10)}$$

In general terms any number (N) can be expressed by the following general equation:

$$N = d_n R^n + d_{n-1} R^{n-1} + \cdot \cdot \cdot \cdot \cdot + d_2 R^2 + d_1 R^1 + d_0 R^0$$

Where N is the number,

d_n is the digit of nth position and
R^n is the base or radix of the nth position

Since computers are simple electronic devices that can only sense voltage on (1) or off (0), a light being on (1) or off (0), a transistor on (1) or off (0), or magnetic field on (1) or off (0), they cannot work with the decimal system's complexity.

Binary Number System

A numbering system that is made up of only the two basic digits "0" and "1" is called base 2 or binary number system. This is the basic system that computers work with; it is also the basis for our punched tape codes.

Comparing the decimal and binary bases with their powers, we find no difference in the principle:

$$10^0 = 1 \qquad 2^0 = 1$$
$$10^1 = 10 \qquad 2^1 = 2$$
$$10^2 = 100 \qquad 2^2 = 4$$
$$10^3 = 1000 \qquad 2^3 = 8$$

The latter is the base our punched tape codes of numerics operate on. The binary numbers can now be handled by on-off type of electronic circuits. The equivalence between decimal and binary numbers is shown in Table 2.1.

TABLE 2-1 DECIMAL AND BINARY NUMBERS			
Decimal	Binary	Decimal	Binary
0	0000	11	1011
1	0001	12	1100
2	0010	13	1101
3	0011	14	1110
4	0100	15	1111
5	0101	16	10000
6	0110	17	10001
7	0111	18	10010
8	1000	19	10011
9	1001	20	10100
10	1010		

Converting Decimal to Binary

EXAMPLE

Convert Decimal 327 to Binary.

Solution: *Remainder:*

2⎣327	1 (LSD) least significant digit
2⎣163*	1
2⎣81*	1
2⎣40*	0
2⎣20*	0
2⎣10*	0
2⎣5*	1
2⎣2*	0
2⎣1*	1 (MSD) most significant digit

Read from the MSD to the LSD, the binary equivalent of 327 is 101000111.

NOTE: (* Quotients)

EXAMPLE

Convert Decimal 92 to Binary.

Solution: *Remainder:*

2⎣92	0 LSD
2⎣46	0
2⎣23	1
2⎣11	1
2⎣5	1
2⎣2	0
1	1 MSD

Read from the MSD to the LSD, the binary equivalent of 92 is 1011100.

EXERCISE

Convert the following decimal numbers to binary:

      a. 23 b. 69 c. 147 d. 1897

Converting Binary to Decimal

Changing the base from 10 to 2 in the general equation discussed under the decimal section is a very simple operation.

EXAMPLE

Determine the decimal value of the following binary numbers:

    1. $(1011)_2 = (?)_{10}$

Solution:

- Assign powers 0, 1, 2, 3, to the binary numbers from right to left (these powers are for the base).

$$1^3 \quad 0^2 \quad 1^1 \quad 1^0$$

- Substitute these binary numbers into the general equation using the base or radix 2 instead of 10, at the power corresponding to the location of the digit, as shown above, and multiply each one by the corresponding binary digit.

$$N = 1 \times 2^3 + 0 \times 2^2 + 1 \times 2^1 + 1 \times 2^0$$
$$= 8 + 0 + 2 + 1 = (11)_{10}$$

2. $(101000111)_2 = (?)_{10}$

Solution:

- $1^8 0^7 1^6 0^5 0^4 0^3 1^2 1^1 1^0$

$$N = 1 \times 2^8 + 0 \times 2^7 + 1 \times 2^6 + 0 \times 2^5 + 0 \times 2^4$$
$$+ 0 \times 2^3 + 1 \times 2^2 + 1 \times 2^1 + 1 \times 2^0$$

- Canceling terms with zero digits

$$N = 1 \times 2^8 + 1 \times 2^6 + 1 \times 2^2 + 1 \times 2^1 + 1 \times 2^0$$
$$= 256 + 64 + 4 + 2 + 1 = (327)_{10}$$

EXERCISE

Determine the decimal values of the following binary numbers:

a. $(10101011)_2$ b. $(1001111)_2$
$= (\quad)_{10}$ $= (\quad)_{10}$

c. $(1011)_2$ d. $(11111101)_2$
$= (\quad)_{10}$ $= (\quad)_{10}$

Fractional Binary Numbers

Since most programmed numbers are fractions of a whole (decimal), fractional numbers are as important as integers. The method of converting fractions to binary numbers differs from the integer method. Instead of dividing, we multiply the fraction by 2. The number to the left of the decimal point of the product will be the binary number, while the sum to the right of the decimal point will be used as multiplicand. This multiplication is repeated until the desired accuracy is attained.

EXAMPLE

$$(0.375)_{10} \quad \rightarrow .750 \quad \rightarrow .500$$
$$\underline{\times \quad 2} \quad \underline{\times \quad 2} \quad \underline{\times \quad 2}$$
$$0.750 \quad 1.500 \quad 1.000$$

Binary 0   1   1

MSD is 0.

The answer is $(0.011)_2$.

Converting fractional binary to decimal is identical to the integer conversion. The general formula we have used for integers used positive powers of the base. For fractions, these powers will have to be changed to negative as:

$$N = d_1 \times R^{-1} + d_2 \times R^{-2} + d_3 \times R^{-3} + \cdots \cdots + d_n \times R^{-n}$$

EXAMPLE

$$(0.011)_2 = (\; ? \;)_{10}$$

Solution:

- Assign powers,

$$0^{-1}1^{-2}1^{-3}$$

These powers, as in our previous example, are transferred to the base as:

$$N = 0 \times 2^{-1} + 1 \times 2^{-2} + 1 \times 2^{-3}$$

Canceling the zero terms and performing the summation as shown below:

$$1 \times 2^{-2} = 1 \times \frac{1}{2^2} = \frac{1}{4} = 0.25$$

$$1 \times 2^{-3} = 1 \times \frac{1}{2^3} = \frac{1}{8} = 0.125$$

$$\therefore \quad 0.25 + 0.125 = (0.375)_{10}$$

Mixed numbers can convert just as easily from binary to decimal if we remember that the powers of the base are positive to the left and negative to the right of the decimal point.

EXAMPLE

Convert $(1011.1011)_2$ to decimal.

Solution:

$$1^3 0^2 1^1 1^0 . 1^{-1} 0^{-2} 1^{-3} 1^{-4}$$
$$N = 1 \times 2^3 + 0 \times 2^2 + 1 \times 2^1 + 1 \times 2^0 + 1 \times 2^{-1} + 0 \times 2^{-2}$$
$$+ 1 \times 2^{-3} + 1 \times 2^{-4}$$
$$= 1 \times 2^3 + 1 \times 2^1 + 1 \times 2^0 + 1 \times 2^{-1} + 1 \times 2^{-3} + 1 \times 2^{-4}$$
$$= 8 + 2 + 1 + 0.5 + 0.125 + 0.0625$$
$$= (11.6875)_{10}$$

EXERCISES

Convert the following numbers:

    a. $(0.6565)_{10} = (\ ? \ )_2$ b. $(0.8759)_{10} = (\ ? \ )_2$

    c. $(0.111011)_2 = (\ ? \ )_{10}$ d. $(0.110101)_2 = (\ ? \ )_{10}$

Mixed numbers will convert to binary in a two-step procedure. First, we convert the integer digits (using the division by 2 process), then we convert the decimal digits (using the multiplication by 2 process).

EXAMPLE

Convert $(29.1875)_{10}$ into binary.

Solution:

· Convert integers:

2⎸29	1 ← LSD
2⎸14	0
2⎸7	1
2⎸3	1
2⎸1	1 MSD Integer answer is 11101

· Convert the decimal digits as:

0.1875	→.3750	→.7500	→.5000
× 2	× 2	× 2	× 2
0.3750┘	0.7500┘	1.5000┘	1.0000
↓	↓	↓	
0	0	1	1 Decimal answer is 0.0011

∴ $(29.1875)_{10} = (11101.0011)_2$

EXERCISES

Convert the following numbers:

    a. $(6.875)_{10} = (\ ? \ )_2$ b. $(0.6985)_{10} = (\ ? \ )_2$

c. $(13.4315)_{10} = (?)_2$ d. $(22.555)_{10} = (?)_2$

There are other number systems, used in computer technology, such as the base 8 $(N)_8$ called octal, and the base 16 $(N)_{16}$ or hexadecimal, in addition to the binary number system. The hexadecimal system is used by IBM360, 370, Honeywell 200, and RCA Spectra 10, as well as by some microcomputers. Detailed discussion of other number systems would be beyond the scope of our objectives.

However, a brief discussion of the four basic arithmetic operations with binary numbers will be a useful aid for the reader (programmer). In each operation, the reader must "memorize" only four combinations, as opposed to the decimal system where one had to "memorize" 100 combinations. For addition in binary numbers, see Table 2-2; Table 2-3 shows subtraction in binary numbers.

Addition

TABLE 2-2 ADDITION IN BINARY NUMBERS			
Augend	Addend	Sum	Carry
0 +	0 =	0	0
1 +	0 =	1	0
0 +	1 =	1	0
1 +	1 =	0	1

EXAMPLE

```
              1  1  1  1  1    ← carries      Note:    1
  A             1  1  1  1              15            + 1
  B    +     1  1  1  1  1            + 31           0, carry 1
A + B     1  0  1  1  1  0           (46)₁₀
```

Subtraction

TABLE 2-3 SUBTRACTION IN BINARY NUMBERS			
Minuend	Subtrahend	Difference	Borrow
0 −	1 =	1	1
1 −	1 =	0	0
1 −	0 =	1	0
0 −	0 =	0	0

EXAMPLE

$$
\begin{array}{rll}
\text{Borrows} & 000000 & \\
A & 11001 & 49 \\
B & -10000 & -32 \\
\hline
A - B & 010001 & (17)_{10}
\end{array}
$$

Subtraction of large binary numbers is difficult, especially for those who are not familiar with the binary number system. Some of you may find it easier to work with another method called Binary Complement Subtraction. The method is described below, step by step, using the same example as above.

1. A   49   1 1 0 0 0 1 ⟶ 1 1 0 0 0 1
   B − 32 − 1 0 0 0 0 0   + 0 1 1 1 1 1
   ‾‾17

2. Write down the minuend as shown on the right. Then write down the complement of the subtrahend below.

3. Instead of subtracting, add the two numbers. Once the addition is completed, the carry must be added to the sum as shown:

$$
\begin{array}{cccccc}
1 & 1 & 1 & 1 & 1 & 1 \\
\end{array} \quad \text{carry}
$$

```
  1  1  1  1  1  1        carry
|  1  1  0  0  0  1              49
|  0  1  1  1  1  1           - 32    Complement of 32
|  0  1  0  0  0  0           ‾‾‾‾‾
↓
(1)——————————→1
   1  0  0  0  1              (17)₁₀
```

NOTE: (The complement of binary 0 is 1 and the complement of binary 1 is 0.)

EXERCISE

Subtract the following numbers:

a. $\left(\begin{array}{r} 50 \\ -48 \\ \hline 02 \end{array}\right)_{10}$ $\begin{array}{r} 110010 \\ - 110000 \\ \hline \end{array}$

b. $\left(\begin{array}{r} 47 \\ -11 \\ \hline 36 \end{array}\right)_{10}$ $\begin{array}{r} 101111 \\ - 1011 \\ \hline \end{array}$

Binary Multiplication

Most CNC systems do not perform multiplications. The multiplication is implemented by repeated addition, the same way as the addition of all partial products is performed to obtain the final sum (see Table 2-4). The formation of partial products is easy (the same as the decimal multiplication). The addition of all the partial product is more difficult. You must count the number of ones in the column: If it is even, the sum of the column is 0. If it is odd, the sum of the column is 1. For every pair of 1s there is one carry to the next higher position.

TABLE 2-4
MULTIPLICATION IN BINARY NUMBERS

Multiplicand		Multiplier		Product
1	*	1	=	1
0	*	1	=	0
1	*	0	=	0
0	*	0	=	0

EXAMPLE

```
            43  A  1 0 1 0 1 1
          × 22  B   ×1 0 1 1 0
            946

          0 1 1 1 1 1 1 0 0    carry
                 *       *     multiply by "0"
Solution:        0 0 0  0 0 0  start with LSD
              1 0 1 0  1 *1    —multiply by 1
            1 0 1 0 *1 1       —multiply by 1
          0 0 *0 0 *0 0        —multiply by 0
        1 0 1 0 1 1            —multiply by 1
   A · B = 1 1 1 0 1 1 0 0 1 0  add
```

NOTE: $\frac{1}{1}$)* = 0 with carry of 1.

The only possible way to verify a large binary number such as our result is through binary-to-decimal conversion (discussed in the preceding pages).

$N = 0 \times 2^0 + 1 \times 2^1 + 0 \times 2^2 + 0 \times 2^3 + 1 \times 2^4 + 1 \times 2^5 + 0 \times 2^6$
$\quad + 1 \times 2^7 + 1 \times 2^8 + 1 \times 2^9$
$= 0 + 2 + 0 + 0 + 16 + 32 + 0 + 128 + 256 + 512$
$= (946)_{10}$

EXERCISE

Perform the following multiplications:

<table>
<tr><td>a.</td><td>11111
× 1110</td><td>b.</td><td>10 1101
x 1011011</td></tr>
</table>

Binary Division

As in the case of multiplication, most CNC systems perform divisions by repeated subtraction of the divisor from the dividend.

The division rules for two 1-bit binary numbers are shown in Table 2-5.

TABLE 2-5 DIVISION IN BINARY NUMBERS			
Dividend	**Divisor**	**Quotient**	**Remainder**
1 ÷ 1		1	0
1 ÷ 0		Undefined	Undefined
0 ÷ 1		0	1
0 ÷ 0		Undefined	Undefined

EXAMPLE

Dividend Divisor Quotient:

$$26 \div 5 = 5.2$$
$$10$$
$$0$$

Solution: 11010 ÷ 101 = 101.001
       101
       ‾‾‾‾‾
       00110
        101
        ‾‾‾
        001000
         101
         ‾‾‾
          11

EXERCISE

a. 11011 ÷ 1011 = (?)   b. 1011 ÷ 111 = (?)

2.1.2   *Binary-Coded-Decimal Code (BCD)*

We have earlier established that the CNC system is an electronic device that can understand simple "on" (1) or "off" (0) states. We have also showed that the base 2 or binary number system allows us to represent any decimal number in binary. All CNC systems use some sort of binary system for their arithmetic or internal operation, but externally the real world works with the decimal system. We have seen that conversion between decimal and binary can be long and erroneous for large numbers. As a compromise, a binary-coded-decimal (BCD) coding was developed, based on the position of the numbers used to describe our CNC tape codes. The first, second, third, and fourth positions can be described as:

$$2^0 = 1, \ 2^1 = 2, \ 2^2 = 4 \text{ and } 2^3 = 8$$

Reading from right to left, the weight can be written as 8-4-2-1. For this reason, this system is often called the 8421 code. This code compresses the binary numbers so that they can be punched in tape to control our CNC system. The BCD (8421) codes are punched in rows across the 1-inch-wide standard tape, each row represents one digit in the tape and successive rows can represent any numbers. Fig. 2-1 illustrates the BCD tape code principle.

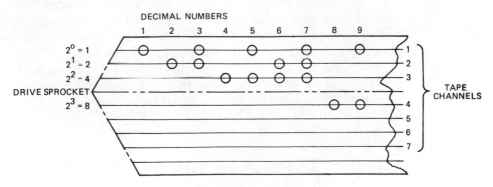

FIG. 2-1. BCD Tape Code.

One of the main advantages of the BCD system is that once learned it is easy to read the values represented by punched holes. For example:

Digit 1 is represented by a hole in channel 1 $(2^0 = 1)$
Digit 2 is represented by a hole in channel 2 $(2^1 = 2)$
Digit 5 has no code of its own, but is the sum of $2^2 = 4$ plus
$2^0 = 1$ hence $4 + 1 = 5$.

The reader can easily visualize the simplicity. The BCD codes are used in both Electronic Industries Association (EIA) and the American Standard Code for Information Interchange (ASCII) Systems, which will be discussed in more detail in the following pages.

2.2   CNC TAPES

When punched or recorded (in the case of magnetic tape), these tapes are predominantly used by CNC systems as input/output or control media. The reader may find many different makes and colors of 1-inch-tape commercially available. All tapes are manufactured to an EIA standard (shown in Fig. 2-2), which also outlines the tolerances required by the manufacturers of tape punching and reading equipment.

Selection of the tape material should be based on the type of tape reader and tape punching unit available.

For mechanical tape readers (few in new systems) mylar base tapes should be used as mylar provides considerably longer life under repeated rereading than the regular paper tape. Although it is more expensive, it does not require special punching facilities, has low wear rate, and provides excellent resistance to oil and grease.

For photo electric tape readers and systems with memory, the inexpensive paper tape should suffice. The primary requirement is to provide high opacity and low reflectivity. However, tape readers with 500 characters per second (cps) or higher reading speeds may damage this tape if repeated readings are required. Because of the high acceleration and deceleration rates, these tape readers are most reliable with laminated paper-mylar or a less costly aluminized-mylar tape.

2.3   TAPE-PUNCHING FACILITIES

Several different manual tape punches are commercially available. However, with the drastic price reductions in the minicomputer field we recommend the purchase of a minicomputer base tape-punching facility that works with a floppy disk drive. The writers are using a TRS-80 system coupled with a tape punch and portable hard copy printer. This system provides an inexpensive trouble-free high-speed tape punch; the program is written on the screen and can be edited (corrected) before a tape is prepared. The diskette provides the lowest cost storage device available as dozens of programs can be recorded on a single diskette. The TRS-80 mini-computer can also be connected to a standard RS-232 teletype terminal for tape punching and hard copy print. Using the teletype as an input/output (I/O) device, the TRS-80 or any other minicomputer having a CRT can easily be used with a minimal software cost to prepare input tape.

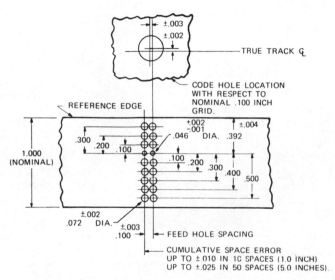

FIG. 2-2. Electronic Industries Association Standards for 1-inch-perforated tape. (Courtesy of Electronic Industries Association)

For computer-aided programming, we should mention that editing software can be purchased with the hardware for most commercially available minicomputers.

CNC systems use two different punched-hole codings: the EIA RS-244 standard was developed and used predominantly by the North American industry; ASCII RS-358 standard was developed in the United States, but is accepted and used throughout the world under the name ISO. While some older NC controls work with EIA, most current CNC systems accept both punched codes.

PARITY CODE: To minimize the possibilities of errors during the internal handling of binary data within the CNC control, as well as during tape punching and reading, a Parity-Check system has been implemented in the standardized coding.

The Parity Check Code for the EIA consists of an extra hole in the fifth row, in addition to the BCD code if the holes would otherwise be even, so

CHARACTER	MAG TAPE CODING							(ISO) ASCII									EIA								
	1	2	3	4	5	6	7	1	2	3	4	5	6	7	8	1	2	3	4	5	6	7	8
0																							
1																							
2																							
3																							
4																							
5																							
6																							
7																							
8																							
9																							
a																							
b																							
c																							
d																							
e																							
f																							
g																							
h																							
i																							
j																							
k																							
l																							
m																							
n																							
o																							
p																							
q																							
r																							
s																							
t																							
u																							
v																							
w																							
x																							
y																							
z																							
+																							
—																							
/																							
.																							
,																							
%																							
&																							
(																NOT		ASSIGNED					
)																NOT		ASSIGNED					
:																NOT		ASSIGNED					
CR																							
DELETE																							
SPACE																							
=																							
TAB																							

FIG. 2-3. ASCII (ISO) and EIA punch tape codes used by CNC controls.

that the number of holes across the tape will always be "odd." Holes in track five are used exclusively for this purpose.

The Parity Check Code for the ASCII requires "even" numbers of holes across the tape in every row. Holes in track eight are used exclusively for parity adjustment whenever it is required.

Punched tape codes were also developed for alphabetic and other symbolic keyboard codes by ASCII. Both ASCII and EIA punch tape codes are illustrated by Fig. 2-3.

2.4   MAGNETIC TAPE CODES

Magnetic tape codes are also derived from the BCD system and in most CNC applications they are recorded (instead of punched) on 1/2-inch-wide cartridged magnetic (Mag) tape. The codes or data are recorded in seven parallel channels or tracks (Fig. 2-3), using a character density of 800 to 4,800 per inch. Their reading speeds are expressed in inches per second (ips) or characters per second (cps). This speed is normally 160 ips or 45,000 cps on most CNC controls. The read-write heads may be of one- or two-gap types. The one-gap head is used for either reading or writing, but only one at a time (more popular with CNC); the two-gap type can write a code (bit) and read it back, while the bit is still under the head, for parity check.

Magnetic tape media was used by several U.S. control manufacturers during the late 1960s and 1970s, mostly by Bendix, Thompson-Rand-Wooldridge, and Kearney & Trecker for four and five-axis contouring. Currently, the "Bandit" control is using a mag tape control. This control is more efficient and less troublesome than earlier types. In spite of the high reading characteristics and the lower basic cost of the control, mag tape never gained the wide acceptance that punched tape did. The codes cannot be seen by the naked eye; however, special optical viewing instruments are available on the market to view the recorded codes, as shown by Fig. 2-3. The alpha numeric codes are shown by dashes (-). Interactive CNC controls with mag tape storage are periodically coming to the market.

Regardless of the type of input media we use, the possibility of errors is proportional with the number of times we handle the data. Using punched tape requires that we punch-store-transport and read in the CNC. The only way to eliminate the error possibilities is to transfer the data (CNC program) directly from the computer to the CNC. Several large companies are now in the process (some have already converted) to change to DNC. Experiments are also under way to develop this type of data transfer from minicomputers.

2.5   MATHEMATICS FOR THE PROGRAMMER

The question most asked by persons wishing to learn CNC programming is "What level of mathematics do I need to be able to learn CNC programming?" Unfortunately there is no simple answer. However, it is safe to state that a part programmer for lathe and/or two-axis contour milling should have a working knowledge of coordinate systems, trigonometry, analytical geometry, and cutting forces. While the objective of this book is not to discuss lengthy mathematical derivations, it will provide valuable information for those who wish to review areas of concern.

2.5.1   *Cartesian Coordinate System*

Most of you readers are familiar with the rectangular or cartesian coordinate system you have learned in high school. All the CNC systems are built to function and therefore must be programmed in terms of a coordinate system. The mathematics discussed here will be shown in coordinate systems whenever possible. A two-axis coordinate system is formed by two intersecting straight lines perpendicular to each other (see Fig. 2-4), hereafter called X- and Y-axes.

The sample programs in the book will refer to this two-axis system for positioning and contouring. Drawing a third line, as shown by Fig. 2-4, perpendicular to the plane formed by the X-Y axis through the intersecting point will form a three-axis coordinate system. The intersection

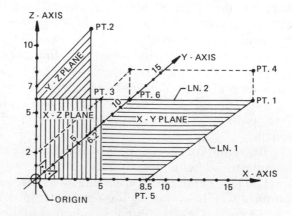

FIG. 2-4. Cartesian coordinated system.

point is called the "Origin" and the third or "Z-" axis will be called the tool axis.

POINT: The simplest element is a point (PT) and it can be defined by its X-Y coordinates as PT1 (X, Y), shown by Fig. 2-4 using actual values as PT1 (8.5, 11). This notation refers to a two-axis coordinate system. PT2 and PT3 cannot be defined by this method because PT2 has no X and PT3 has no Y values. These points must therefore use a three-axis notation in the form of PT (X, Y, Z). This notation allows us to define the location of any point in space in terms of our coordinate system as PT2 (0, 6.2, 7), PT3 (5, 0, 6), and PT4 (8.5, 11, 2).

LINE: Any line can be defined by two points in a cartesian coordinate system. (There is another definition using a radius from the origin and an angle measured from the positive X-axis which will be discussed under Polar Coordinates.)

EXAMPLE

Define Line 1 (LN1) and Line 2 (LN2).

Solution:

PT1 (8.5, 11, 0) is defined in Fig. 2-4.
Define the second point, PT5 as PT5 (8.5, 0, 0)

Now Line 1 can be defined as LN1 (PT1, PT5), or a line going through points PT1 and PT5.

SIMILARLY: LN2 (PT1, PT6), where PT6 (0, 11, 0).

PLANE: Better known in industry as surface, it can be defined by three points. It can also be defined several other ways. However, plane definitions by rotation and transfer are beyond the objective of this book. Some planes, such as PL1, PL2 or PL3 are illustrated in Fig. 2-5.

EXAMPLE

Define Plane 3.

Solution:

Define the three points required for the plane definition.

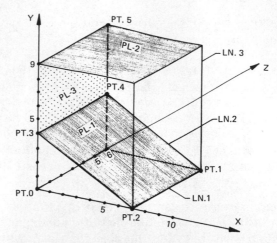

FIG. 2-5. Elements of geometry in a cartesian coordinate system.

PT3 (0, 4, 0), PT4 (0, 4, 6), PT5 (0, 9, 6)

Define the plane as PL3 (PT3, PT5, PT4).

EXERCISE

Define the location of the following geometry:

      a. PT1 (, ,) b. LN1 (,)   c. LN2 (,)

      d. LN3 (,)   e. PL1 (, ,) f. PL2 (, ,)

2.5.2   *Polar Coordinate System*

The nomenclature of the axis (X-Y-Z) is identical to the cartesian system. However, the coordinate location of the point, line, or plane is defined in terms of a radius (distance from origin to a point) and the angle between the positive X-axis and the geometric shape we wish to specify. The angle is positive (+) in the counterclockwise direction (CCW) and negative when measured in the clockwise (CW) direction from the X-axis. See Fig. 2-6.

    Some of the latest CNC controls are programmed in terms of polar coordinates. This simplifies the calculations when holes have to be drilled on a circular pattern.

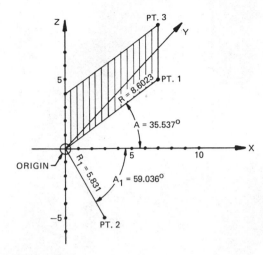

FIG. 2-6. Polar and/or cylindrical coordinate system.

EXAMPLE

Define the location of PT1 in terms of polar coordinates.

Solution:

PT1 (R, A)
PT1 (8.6023, 35.537)

EXERCISE

Define the location of PT2.

PT2 (,)

Points not located in the reference plane are defined by their "cylindrical coordinates." PT3 must therefore be defined in terms of its radius "R," angle "A," and height "Z," in the form of PT3 (R, A, Z). Using the dimensions from Fig. 2-6, the answer will be PT3 (8.6023, 35.537, 4.0).

A typical CNC application of the cylindrical coordinate system is illustrated in Fig. 2-7.

In order to mill the cam groove on the cylinder (centerline of groove shown on drawing), we need first to define the start (PT1) and end (PT2)

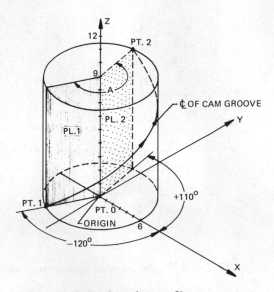

FIG. 2-7. Cylindrical coordinate system.

points in terms of the radius, angle, and height dimensions. The tool path from − 120° to + 110° describes the rotation of plane 1 to plane 2 position. The rotation plane X-Y is circular, and the third axis motion (Z) is linear. In fact, the tool point will describe a helical motion along the surface of a perpendicular cylinder.

EXERCISE

Define the location of PT1 and PT2 in the cylindrical coordinate system.

PT1 (, ,)   PT2 (, ,)

2.6   TRIGONOMETRIC FUNCTIONS

The science of "triangle measurement" is commonly known as "trigonometry." Trigonometric functions such as angles, sides of right-angle triangles, and their relationships will be discussed in this section.

2.6.1   Pythagorean Theorem

The square of the hypotenuse in a right-angle triangle is equal to the sum of the squares of the other two sides. See Fig. 2-8.

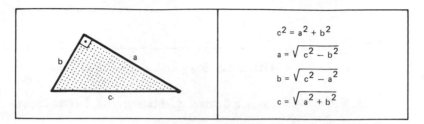

FIG. 2-8. Pythagorean Theorem.

2.6.2   Similar Triangles

If the sides of any angle are intersected by two parallel straight lines, two similar triangles are formed. The ratios of the sides can be expressed as shown by Fig. 2-9. This relation will hold for any number of parallel lines traced, i.e., a2, b2, c2−a3, b3, c3, etc.

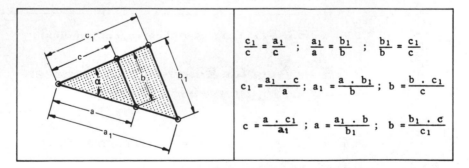

FIG. 2-9. Similar triangles.

2.6.3   Sine and Cosine Functions

We will only show the most frequently used functions (Fig. 2-10).

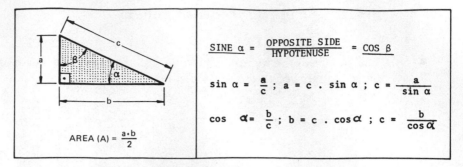

FIG. 2-10. Sine and cosine functions.

2.6.4   *Tangent and Cotangent Functions*

The most frequently used functions are illustrated in Fig. 2-11.

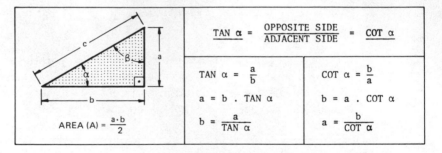

FIG. 2-11. Tangent and cotangent functions.

2.6.5   *Angular Relationships Between Trigonometric Functions*

If $a = c \cdot \sin \alpha$ and $b = c \cdot \cos \alpha$ from 2.6.3 and $\text{Tan } \alpha = \dfrac{a}{b}$ from 2.6.4

then by substitution $\text{Tan } \alpha = \dfrac{c \cdot \sin \alpha}{c \cdot \cos \alpha}$

or $\boxed{\text{Tan } \alpha = \dfrac{\sin \alpha}{\cos \alpha}}$; $\text{Sin } \alpha = \text{Tan } \alpha \cdot \cos \alpha$ and

$$\text{Cos } \alpha = \frac{\sin \alpha}{\tan \alpha}$$

Similarly

$$\text{Tan } \alpha = \frac{a}{b} \quad \text{and} \quad b = a \cdot \cot \alpha \quad \text{by substitution}$$

$$\text{Tan } \alpha = \frac{a}{a \cdot \cot \alpha} \quad \text{therefore} \quad \boxed{\text{Tan } \alpha = \frac{1}{\cot \alpha}}$$

$$\text{if Tan } \alpha = \frac{\sin \alpha}{\cos \alpha} = \frac{1}{\cot \alpha}$$

then . . . . $\boxed{\text{Cot } \alpha = \frac{\cos \alpha}{\sin \alpha}}$

2.7   OBLIQUE TRIANGLES

Sometimes the programmer has to do calculations of angles or sides of triangles that do not have a 90° angle. Some calculation procedures for oblique triangles are shown by Fig. 2-12.

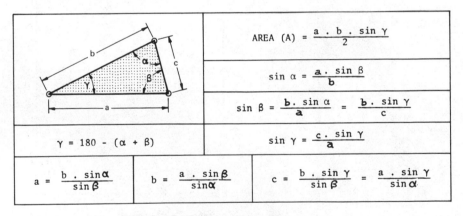

	$\text{AREA (A)} = \dfrac{a \cdot b \cdot \sin \gamma}{2}$
	$\sin \alpha = \dfrac{a \cdot \sin \beta}{b}$
	$\sin \beta = \dfrac{b \cdot \sin \alpha}{a} = \dfrac{b \cdot \sin \gamma}{c}$
$\gamma = 180 - (\alpha + \beta)$	$\sin \gamma = \dfrac{c \cdot \sin \gamma}{a}$
$a = \dfrac{b \cdot \sin \alpha}{\sin \beta} \qquad b = \dfrac{a \cdot \sin \beta}{\sin \alpha}$	$c = \dfrac{b \cdot \sin \gamma}{\sin \beta} = \dfrac{a \cdot \sin \gamma}{\sin \alpha}$

FIG. 2-12. Oblique triangles.

2.8   ANALYTIC GEOMETRY

Analytic geometry is the science that deals with the graphical representation of an equation. We are mainly interested in introducing the reader to points, lines, and circles, their intersections and relationships in a coordinate system. The reason behind this interest is that outside and inside contours of most parts machined on CNC equipment can be defined in terms of lines and circles. Programmers interested in studying more complex curves such as parabola, ellipse, hyperbola, and others, as well as three-dimensional analytic geometry, will find specialized texts dealing exclusively with this topic.

2.8.1 *Equation of a Straight Line*

A line may be defined through its Y-intercept (the point at which it intersects the Y-axis) and its slope in relation to the positive X-axis. See Fig. 2-13. The slope-intercept equation can be written as:

$$y = m \cdot x + b$$

Where $m = \tan \alpha$ and
$b = 8$

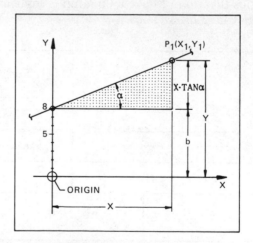

FIG. 2-13. Linear graph.

EXAMPLE

1. Find the "y" dimensions for

$$X = 6.0 \quad \text{and} \quad X = 8.0$$
$$\text{if} \quad \alpha = 30° \quad \text{and} \quad b = 8$$

Solution:

Both points are located on the above line. Their coordinate axes must therefore satisfy the requirements of the equation.

Tan 30° = 0.57735; this is the slope "m."

Inserting the values of m and b, we obtain the equation of the line:

$$y = 0.57735 \cdot x + 8$$

$$\text{For} \quad x = 6 \quad y = 0.57735 \cdot 6 + 8 = 11.4641 \quad \text{and}$$
$$\text{For} \quad x = 8 \quad y = 0.57735 \cdot 8 + 8 = 12.6188$$

If, on the other hand, we know the coordinates (x_1, y_1) of a point P_1 and the slope of the line, its equation can be obtained from the following formula:

$$y = y_1 + m \cdot (x - x_1)$$

2. Find the slope-intercept form of the equation of a line defined by an angle $\alpha = 30°$ and passing through a point P_1 of coordinates $x_1 = 6.0$ and $y_1 = 11.4641$.

Solution:

$$\text{Tan } 30° = 0.57735 = m$$
$$y = 11.4641 + 0.57735 \cdot (x - 6.0)$$
$$y = 11.4641 + 0.57735 \cdot x - 3.4641 = 0.57735 \, x + 8.0$$
$$y = 0.57735 \cdot x + 8.0$$

which is the slope intercept equation used at the start of this paragraph.

Using the above equations, any y-coordinate can be found in terms of its x-coordinate.

2.8.2   Equation of a Circle

If the center of a circle is at the origin of the coordinate system, the equation of the circle is:

$$r^2 = x^2 + y^2$$

If the center of the circle is not in the origin of the coordinate system, but located in a point $Q(X_Q, Y_Q)$, the equation of the circle will be:

$$r^2 = (x - x_Q)^2 + (y - y_Q)^2$$

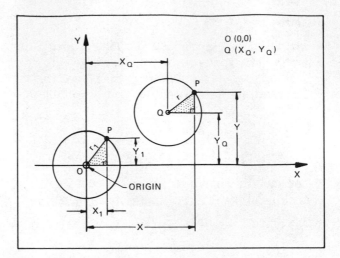

FIG. 2-14. Circular graph.

Applying these basic principles, the programmer wil be able to calculate the intersection point coordinates for line-line, line-circle, and circle-circle relationships.

2.8.3 Intersection of Two Lines

EXAMPLE

Find the intersection point of the following two lines, given in slope-intercept form:

$$\text{LN1} \quad Y = 0.5x + 3.25$$
$$\text{LN2} \quad Y = -2.3x + 7$$

If m_1 and m_2 are the slopes of the two lines, and b_1 and b_2 their respective y-intercepts, the coordinates x_p and y_p of the intersection point PT1 (see Fig. 2-15) can be calculated as follows:

$$x_p = \frac{b_2 - b_1}{m_1 - m_2}$$

$$y_p = m_1 \frac{b_2 - b_1}{m_1 - m_2} + b_1$$

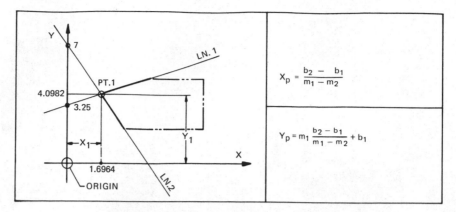

FIG. 2-15. Intersection of two lines.

Solution:

$$X_p = \frac{b_2 - b_1}{m_1 - m_2} = \frac{7 - 3.25}{0.5 - (-2.3)} = \frac{4.75}{2.8} = 1.6964$$

Substitute this value of x_p into the equation of LN1 as follows:

$$y_p = 0.5 \cdot 1.6964 + 3.25 = 4.0982$$

EXERCISE

Find the coordinates X and Y of the point PT1 of intersection of the following lines:

1. LN1 $y = 0.75x + 4.6$ LN2 $y = -3x + 14$

2. LN3 $y = 3x - 6$ LN4 $5y = -2x + 10$

3. Graph all 4 lines in one coordinate system.

2.8.4   Intersection of a Circle and a Line

The coordinates of both points of intersection between the line and circle must satisfy both their equations (see Fig. 2-16). Therefore, the equation of the line is equal to the equation of the circle as follows:

$$Y_p = m \cdot X_p + b = Y_Q \pm \sqrt{r^2 - (X - X_Q)^2}$$

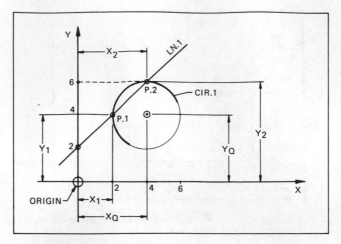

FIG. 2-16. Intersection of line and circle.

NOTE: $Y_Q \pm \sqrt{r^2 - (X - X_Q)^2}$ *was derived from the equation of the circle as shown:*

$$(Y - Y_Q)^2 = r^2 - (X - X_Q)^2$$
$$Y - Y_Q = \pm \sqrt{r^2 - (X - X_Q)^2}$$
$$Y = Y_Q \pm \sqrt{r^2 - (X - X_Q)^2}$$

EXAMPLE

Find the coordinates (X_1, X_2, Y_1, Y_2) of the intersection points (PT1, PT2) from the system of equations of:

Line (LN1) $y = x + 2$
Circle (CIR1) $y = 4 \pm \sqrt{2^2 - (x - 4)^2}$

Solution:

$$y = x + 2 = 4 \pm \sqrt{2^2 - (x - 4)^2}$$
$$(x + 2 - 4)^2 = 4 - (x^2 - 8x + 16)$$
$$(x + 2 - 4)^2 = -x^2 + 8x - 12$$
$$x^2 - 6x + 8 = 0$$

or

$$x = \frac{6 + \sqrt{36 - 32}}{2} = 3 \pm 1$$

$x_1 = 3 + 1 = 4, x_2 = 3 - 1 = 2$
$y_1 = 4 + 2 = 6, y_2 = 2 + 2 = 4$
and the answers are PT1 (4, 6) and PT2 (2, 4).

2.8.5   *Intersection of Two Circles*

EXAMPLE

Find the intersection points of the following two circles:

Circle 1 of equation $y = \sqrt{r^2 - x^2}$ for $r = 4$ and

Circle 2 of equation $y = y_Q \pm \sqrt{r^2 - (x - x_Q)^2}$ for $x_Q = y_Q = 3$
                                                                                                and $r = 3$

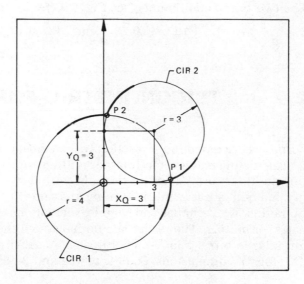

FIG. 2-17. Intersection of two circles.

Solution:

Since the intersection points are common for both circles, we can equate the two equations as:

$$\sqrt{r^2 - x^2} = Y_Q \pm \sqrt{r^2 - (x - x_Q)^2}$$

Substituting:

$$\sqrt{4^2 - x^2} = 3 \pm \sqrt{3^2 - (x - 3)^2}$$

Squaring both sides:

$$16 - x^2 = 9 \pm 6\sqrt{-x^2 + 6x} + (-x^2 + 6x)$$

Solving the equation:

$$\left(\frac{7}{6} - x\right)^2 = -x^2 + 6x \text{ which becomes } x^2 - 4.16x + 0.68 = 0$$

Solving the quadratic equation for x, we obtain:

$$x = \frac{4.16 \pm \sqrt{4.16^2 - 2.72}}{2} = 2.08 \pm 1.91$$

And the coordinates are:

$$x_1 = 2.08 + 1.91 = 3.99 \quad \text{and} \quad y_1 = \sqrt{16 - 3.99^2} = 0.28$$
$$x_2 = 2.08 - 1.91 = 0.17 \quad\quad\quad y_2 = \sqrt{16 - 0.17^2} = 3.99$$

The two intersection points (see Fig. 2-17) will therefore be:

$$P1\ (3.99,\ 0.28) \quad \text{and} \quad P2\ (0.17,\ 3.99)$$

2.9   TRIGONOMETRIC FORMULAS

Formulas discussed in this section will be most useful to the programmer for calculating cutter centerlines for milling applications and TNR center-line paths for turning applications. In milling, "r_c" will be used to identify end mill radius, while in turning, the same "r_c" will represent the tool tip radius. The formulas and sample calculations are given in X-Y coordinates for milling. The reader should have no difficulty in applying these formulas to turning in X-Z coordinates by substituting "Z" for "X" and "X" for "Y" dimensions. See Fig. 2-18a and 2-18b.

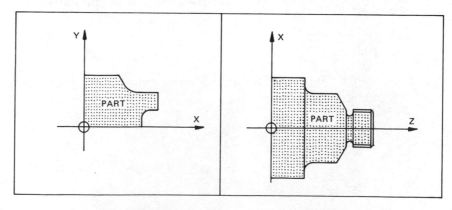

FIG. 2-18. a) Milling.   **b) Turning.**

2.9.1 Cutter Centerline Intersection Point of a Line Parallel with the X-Axis (Z for Lathe) and a Line at an Angle Measured from the First Line

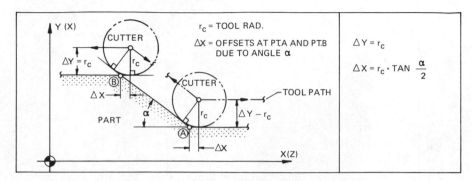

FIG. 2-19. Line parallel with the X-axis.

2.9.2 Cutter Centerline Intersection Point of a Line Parallel to the Y-Axis (X for Lathe) and a Line at an Angle Measured from the X-Axis

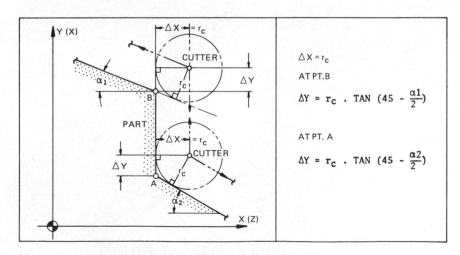

FIG. 2-20. Line parallel with the Y-axis.

2.9.3 Cutter Centerline Intersection Point of Two Lines

Neither line is parallel with the primary (X-Y) axes of the part or machine coordinate systems. See Fig. 2-21.

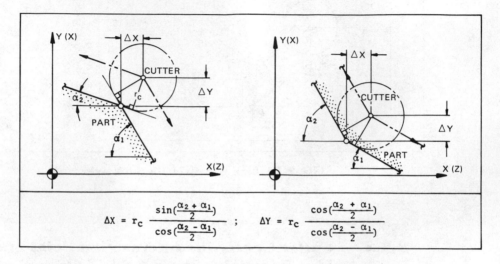

$$\Delta X = r_c \ \frac{\sin(\frac{\alpha_2 + \alpha_1}{2})}{\cos(\frac{\alpha_2 - \alpha_1}{2})} \quad ; \quad \Delta Y = r_c \ \frac{\cos(\frac{\alpha_2 + \alpha_1}{2})}{\cos(\frac{\alpha_2 - \alpha_1}{2})}$$

FIG. 2-21. Lines not parallel with primary axes.

2.9.4 *Cutter Centerline Intersection Point of a Line and a Circle*

The line tangent to the circle, not parallel to either X- or Y-axis of the coordinate system. See Fig. 2-22.

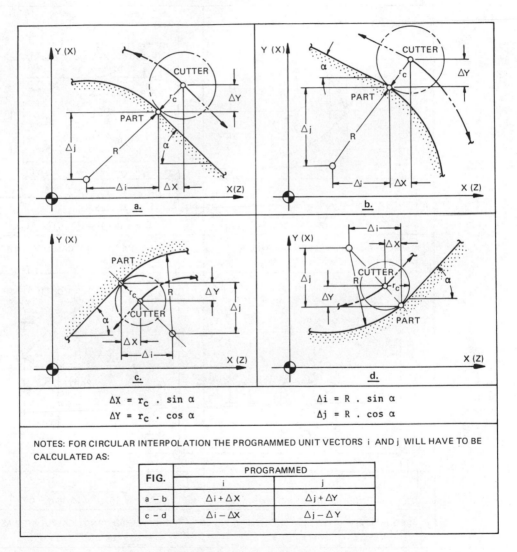

$$\Delta X = r_c \cdot \sin \alpha$$
$$\Delta Y = r_c \cdot \cos \alpha$$

$$\Delta i = R \cdot \sin \alpha$$
$$\Delta j = R \cdot \cos \alpha$$

NOTES: FOR CIRCULAR INTERPOLATION THE PROGRAMMED UNIT VECTORS i AND j WILL HAVE TO BE CALCULATED AS:

FIG.	PROGRAMMED	
	i	j
a – b	$\Delta i + \Delta X$	$\Delta j + \Delta Y$
c – d	$\Delta i - \Delta X$	$\Delta j - \Delta Y$

FIG. 2-22. Line tangent to circle.

2.9.5 Cutter Centerline Intersection Point of a Circle and a Line

The lines parallel to the X- or Y-axis, intersecting a circle. See Fig. 2-23.

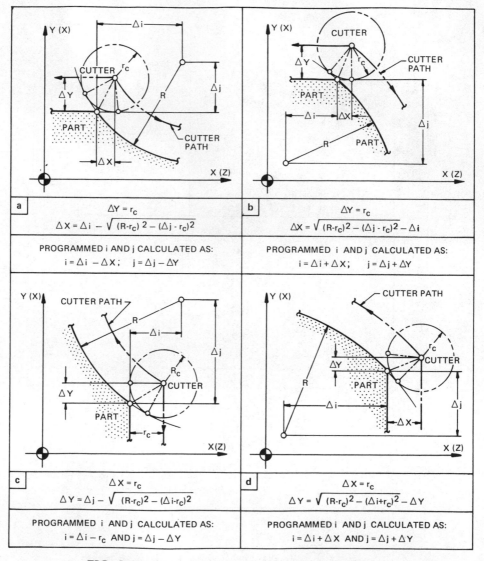

a

$$\Delta Y = r_c$$

$$\Delta X = \Delta i - \sqrt{(R\text{-}r_c)^2 - (\Delta j - r_c)^2}$$

PROGRAMMED i AND j CALCULATED AS:

$$i = \Delta i - \Delta X; \quad j = \Delta j - \Delta Y$$

b

$$\Delta Y = r_c$$

$$\Delta X = \sqrt{(R\text{-}r_c)^2 - (\Delta j - r_c)^2} - \Delta i$$

PROGRAMMED i AND j CALCULATED AS:

$$i = \Delta i + \Delta X; \quad j = \Delta j + \Delta Y$$

c

$$\Delta X = r_c$$

$$\Delta Y = \Delta j - \sqrt{(R\text{-}r_c)^2 - (\Delta i \cdot r_c)^2}$$

PROGRAMMED i AND j CALCULATED AS:

$$i = \Delta i - r_c \text{ AND } j = \Delta j - \Delta Y$$

d

$$\Delta X = r_c$$

$$\Delta Y = \sqrt{(R\text{-}r_c)^2 - (\Delta i + r_c)^2} - \Delta Y$$

PROGRAMMED i AND j CALCULATED AS:

$$i = \Delta i + \Delta X \text{ AND } j = \Delta j + \Delta Y$$

FIG. 2-23. Intersection points of lines and circles.

2.9.6   *Cutter Intersection Point of a Line Tangent to Two Circles*

A typical turning application is illustrated by Fig. 2-24.

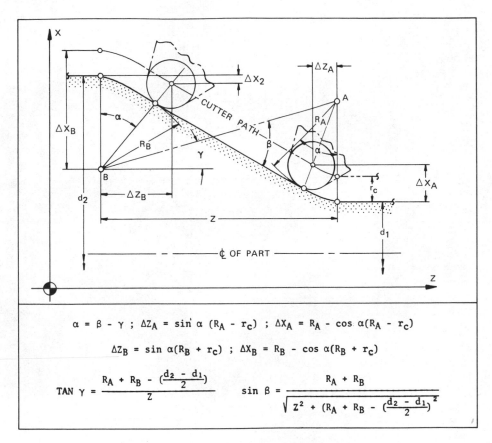

$$\alpha = \beta - \gamma \; ; \; \Delta Z_A = \sin \alpha (R_A - r_c) \; ; \; \Delta X_A = R_A - \cos \alpha (R_A - r_c)$$

$$\Delta Z_B = \sin \alpha (R_B + r_c) \; ; \; \Delta X_B = R_B - \cos \alpha (R_B + r_c)$$

$$TAN \; \gamma = \frac{R_A + R_B - (\frac{d_2 - d_1}{2})}{Z} \qquad \sin \beta = \frac{R_A + R_B}{\sqrt{Z^2 + (R_A + R_B - (\frac{d_2 - d_1}{2}))^2}}$$

FIG. 2-24. Line tangent to two circles.

These formulas can only be useful to the programmer if a working knowledge is gained by solving numerous problems. Readers wishing to expand their mathematical knowledge beyond the scope of this chapter should refer to specialized manuals of analytical geometry.

Computer
Numerical
Control
Systems

In the previous chapter we have discussed how the special-purpose CNC accepts information in the form of punched or magnetic tape codes. This input data must be transformed by the CNC into specific output codes in terms of voltages, or pulses per second (pps). The transformed data, called output, is used to drive the motors to position the machine slides to the programmed position. These slides, or table drives, are commonly known as servodrives. The principal function of the CNC is the positioning of the tool or the machine table in accordance with the programmed data. Industry has developed two distinctly different types of drives based on how the CNC system accomplishes positioning. These are the open-loop and the closed-loop drive systems.

3.1   OPEN-LOOP SERVODRIVES

An automatic washing machine is a typical example of an open-loop system. It will perform a fixed cycle regardless of the state of cleanliness of its contents. In the open-loop servodrive control system, the power supply level is set to a position for which the desired speed is indicated by the input. Should the load on the machine slide vary, the servomotor speed would also be affected. However, the speed variance could not be sensed automatically because the system lacks feedback. For this reason, open-loop control systems can only be used in applications in which there is no change in load conditions. A typical application of the open-loop control is the NC drilling machine. The function of the servo-drive is to position the machine slide or table; therefore, the load condition remains constant. Changes in load due to the weight of the component being machined are taken care of by the design of the machine tool. Manufacturers will normally specify the maximum admissible weight of the parts that can be machined on their product. See Fig. 3-1.

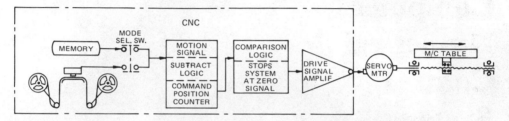

FIG. 3-1. Open-loop CNC positioning control.

In the open-loop servosystem, the motor continues to turn until the absence of power indicates that the programmed location has been attained and the driving mechanism is disengaged. There is no monitoring of this position, and if any movement takes place at this time, its magnitude would normally be unknown. Nevertheless, open-loop control systems have been refined to 0.0001 inch (0.0025 mm) resolution. The systems are reliable, considerably less expensive than closed-loop systems, and their maintenance is far less complicated. Periodical adjustments are required to compensate for wear, as well as deterioration of servodrive components. In summary, this system counts pulses and it cannot identify discrepancies in position.

3.2   CLOSED-LOOP SERVODRIVES

The closed-loop system used in CNC is characterized by the presence of feedback. The term "feedback" is used to describe the various methods of transmitting positional information on the machine slide motion back to the information command section of the CNC. This information is continuously compared with the programmed slide motion data.

CNC systems use two different feedback principles. The indirect feedback monitors the output of the servomotor, as shown in Fig. 3-2a. Although this method is popular with CNC systems, it is not as accurate as direct feedback, which monitors the load condition in the feedback loop, as shown in Fig. 3-2b.

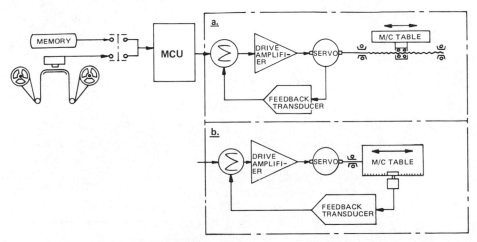

FIG. 3-2. Schematic diagram of feedback systems:
a) Indirect feedback.
b) Direct feedback.

The feedback device is commonly known as "transducer," and may be represented by linear or circular electric scales, shaft digitizers, magnetic scales, or synchros. All these different feedback methods have one thing in common. If the system is digital, the feedback device is counting pulses. If it is analog, the feedback is comparing varying voltage levels. A closed-loop system, regardless of the type of feedback device, will constantly try to achieve and maintain a given position by self-correcting to a zero pulse or a voltage null. The indirect feedback system (Fig. 3-2a) compares the command position signal with the drive signal of the servomotor. This system is unable to sense backlash or leadscrew windup due to varying loads. The direct feedback, with its drive signal originated by the table, is the preferred system because it monitors the actual position of the table on which the part is mounted. The direct feedback system is also called positional feedback system. It is more accurate; however, its implementation costs are higher. While the subject of feedback and servodrives is a science in its own right, a description of a rotary-type transducer used in a digital system (see Fig. 3-3) may illustrate a small portion of a control system.

The rotary disk is attached to the moving part of the machine table. The pinion (disk) is rotated as the table movement takes place, and each

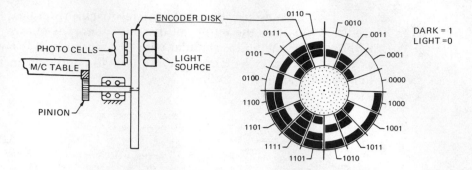

FIG. 3-3. Rotary Encoder for digital transducer.

light interruption from the photo cells creates an electric pulse. If this pulse is the minimum programmable increment, we have a direct one-to-one feedback to the drive command. The encoder is a rotary disk with clear and darkened areas, and it is designed so that for each fixed fraction of a degree, another binary combination is encoded. Fig. 3-3 shows a 4-bit encoder using "Gray-code," a cyclic code generating one change at a time. See Table 3-1. Binary encoders used on earlier systems produced erroneous readings because of ambiguity, or overlap, in reading a code position. The problem was compounded by higher accuracy requirements when 15- or 16-bit resolution was required with a small disk.

A four-resolution segmentation is calculated as follows:

$$360/2^4 = 360/16 = 22.5^0$$

In the previous calculation, the power of 2, i.e., 4, represents the number of rings. The disk, when rotated clockwise, will read the Gray codes as marked. These codes can easily be converted to binary using a simple single flip-flop.

TABLE 3-1 DECIMAL-BINARY-GRAY CODES					
Decimal	Binary	Gray	Decimal	Binary	Gray
0	0000	0000	8	1000	1100
1	0001	0001	9	1001	1101
2	0010	0011	10	1010	1111
3	0011	0010	11	1011	1101
4	0100	0110	12	1100	1010
5	0101	0111	13	1101	1011
6	0110	0101	14	1110	1001
7	0111	0100	15	1111	1000

Usually optical disks containing 2^{10} to 2^{15} bits are used. For $2^{10} = 1024$, the segmentation would be $360/2^{10} = 360/1024 = 0.3515^0$. This requires instrumentation manufacturing techniques for production, and, as a result, the disks become quite expensive.

3.3   VELOCITY FEEDBACK

The positional feedback previously discussed can be implemented to a very high degree of accuracy. However, the system may not provide the required path or surface finish accuracy because no time constraints were provided to reach the programmed or final position. CNC systems used for contouring must have a velocity feedback as well, in order to produce linear, circular, and/or parabolic interpolation, at the same time as acceleration and deceleration velocities.

Feedback is normally provided by an AC or DC tachometer coupled to the servomotor. The feedback of the tachometer is used to modify the positional feedback. The feedback schematics shown in Fig. 3-1 and Fig. 3-2 do not incorporate a velocity loop. Most actual CNC position feedback loops provide some kind of velocity control inside the positioning loop, even if it is not specifically expressed.

The importance of the velocity loop can best be described by the example discussed below.

EXAMPLE

Calculate the displacement inaccuracy of a CNC pulse drive system without a velocity feedback for the following motion:

> N019 G01 X82.55 Y44.45 F100 (in metric), or
> N019 G01 X3.25 Y1.75 F3.94 (in inches), where
>
>> N019 is the sequence number of the tape block
>> G01 is the preparatory code for linear interpolation
>> X82.55 is the programmed displacement or address along X-axis
>> Y44.45 is the programmed displacement or address along Y-axis
>> F100 is the tool velocity, in mm/min in the metric line.

The following assumptions are made:

- The smallest programmable increment is 0.002 mm (0.0001 inches)
- The CNC has pulse servodrives

- To produce the tool path illustrated in Fig. 3-4, both slides must start and stop simultaneously.

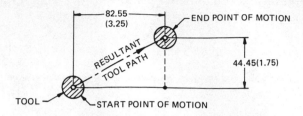

FIG. 3-4. Tool path.

Solution:

1. Calculate the number of pulses required for both the X82.55-mm and Y44.45-mm displacements.

$$\frac{82.55}{0.002} = 41,275 \text{ pulses}$$

$$\frac{44.45}{0.002} = 22,225 \text{ pulses}$$

Since both slides must start and stop simultaneously, these calculated pulses corresponding to the total displacement must be distributed on a different time scale to produce the resultant tool path.

2. Calculate the number of pps required to generate the tool path.

- X time interval

$$\frac{82.55 \text{ mm}}{100 \text{ mm/min}} = 0.8255 \text{ minutes}$$

$$\frac{41,275 \text{ pulses}}{0.8255 \text{ min.}} = 50,000 \text{ ppm}$$

$$= 833.333 \text{ pps}$$

- Y time interval

This is established as a function of the X-displacement, in a direct ratio of the two programmed values for X and Y.

$$\frac{44.45}{82.55} = 0.538462$$

$$0.538462 \cdot 833.333 = 448.718 \text{ pps}$$

3. Calculate the error.

The CNC cannot generate fractional pulses. Therefore, the inaccuracy of the system can be calculated as follows:

- X-error

  0.8255 minutes · 60 = 49.53 seconds motion time
  49.53 · 833 pps = 41,258 pulses (compare with 41,275)
  41,258 pulses · 0.002 mm = 82.5169 mm
  The error = 82.55 − 82.5169 = 0.033 mm or 0.0012 inches.

- Y-error

  49.53 · 448 pps = 22,189 pulses (compare with 22,225)
  22,189 pulses · 0.002 mm = 44.3788 mm
  The error = 44.45 − 44.3788 = 0.071 mm or 0.0027 inches.

In the case of the velocity feedback, the errors can be substantially reduced by applying a larger time scale.

The DC drive systems can work with fractional voltages; therefore, this error would not occur.

Current CNC servosystems are refined to a degree that manufacturers can guarantee 0.002-mm accuracy without difficulty.

EXERCISE

Calculate the displacement inaccuracy of a CNC pulse drive system having no velocity feedback for the following motions:

> N069 G01 X3.756 Y6.925 Z1.25 F6.0 (in inches), and
> N070 G01 X62.05 Y17.372 Z22.111 F221 (in mm.)

The control for circular or parabolic interpolation is a great deal more intricate. The pps must constantly be varied for both axes in order to generate an arc. Readers who wish to study the control aspects of circular or parabolic interpolation should have no difficulty finding texts on the topic of servodrive design.

3.4 POINT-TO-POINT POSITIONING CONTROL

The principal function of the point-to-point positioning control is to position the tool from one point to another within a coordinate system, therefore, the control is most often referred to as a point-to-point NC system.

The positioning may be linear in the X-Y plane, or linear and rotary if the machine has a rotary table. Each tool axis is controlled independently; therefore, the programmed motion may be simultaneous or sequential, but always in rapid traverse. Machining can only take place after positioning is completed. The most common applications of the point-to-point control are in drilling, boring, tapping, riveting, pipe bending, and sheet metal punching.

In addition to positioning, this system is capable of controlling auxiliary functions such as tool change, spindle and coolant on and off, part or fixture clamping, indexing, etc. These on/off type of relay functions do not require intricate control logic, nor do they have any velocity (feed) control.

The tool path of point-to-point control is illustrated in Fig. 3-5.

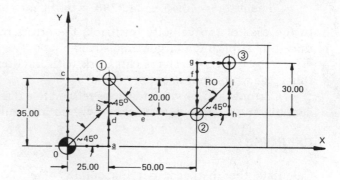

FIG. 3-5. Tool path of point-to-point control.

If the positioning is sequential, the system will move in one axis at a time (as illustrated by the dotted lines). If the positioning is simultaneous, both axes start to move at the same time. Assuming that both drives have the same speed, the tool path will be approximately 45° to the point where the lesser of the two dimensions was completed (as illustrated by solid

	Tool Path		Motion	
TABLE 3-2				
PATHS OF THREE DRILLED HOLES				
Programmed	*Sequential*	*Simultaneous*	*From*	*To*
X25.00 Y35.00	0-a-1 or 0-c-1	0-b-1	0	1
X50.00 Y-20.00	1-d-2 or 1-f-2	1-e-2	1	2
X20.00 Y30.00	2-h-3 or 2-g-3	2-i-3	2	3

lines). From this intermediary point to the final programmed point, the motion will be parallel with the primary axis of the system. The paths described in Table 3-2 are an additional illustration of the drilling of the three holes shown in Fig. 3-5.

Most point-to-point NC systems are built with open-loop drives.

3.5   STRAIGHT-CUT POSITIONING SYSTEMS

The straight-cut positioning system provides a limited degree of control during the positioning of the tool from one point to another. Most of the straight-cut systems are fitted with manually adjustable feed control. This feed control is shared by all the programmable axes of the NC machine, which allows the system to perform milling, in addition to the drillinglike operations outlined in the previous paragraph. Because of this shared feed control feature, the system can also perform milling operations at 45° to the primary axes of the machine.

Realizing this limitation, the programmer can reduce the programmed steps to small enough increments to mill any straight-line pattern and produce the required surface finish and accuracy for most industrial applications. The accuracy this system can produce would most certainly be sufficient for any rough milling application.

The tool path of the straight-cut system is shown in Fig. 3-6 and by the following example:

EXAMPLE

The coordinates of point P2 are X120.00 mm and Y60.00 mm.

Programming these dimensions would result in a tool path from 0 to 1 to P2, with the following error:

$$\text{Error} = \text{distance } 1a = 01 \cdot \sin \alpha = \sqrt{x_1^2 + y_1^2} \cdot \sin \alpha = \sqrt{2(y_1)^2} \cdot \sin \alpha$$
since $x_1 = y_1$
and $\alpha = 45° - \beta = 45° - 26.565° = 18.435°$
The value of β was obtained from the X-Y coordinates of point P2.

Continuing the calculation,

$$\text{Error} = \sqrt{2 \cdot (y)^2} \cdot \sin 18.435° = \sqrt{2 \cdot (60)^2} \cdot \sin 18.435° =$$
26.8328 mm
This value of the error is obviously too large.

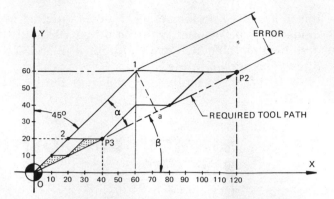

FIG. 3-6. Tool path of straight-cut system.

Reducing the programmed increments to X40.00 mm and Y20.00 mm would result in a tool path of 0 to 2 to P3, and the error would be:

$$\text{Error} = \sqrt{2 \cdot (y)^2} \cdot \sin \alpha = \sqrt{2 \cdot (20)^2} \cdot \sin 18.435° = 8.944 \text{ mm}.$$

This value is proportionally less, but still far too large.

Using this method, we can reduce the programmed increment to steps small enough to yield the desired accuracy.

If the programmed steps are X0.5 mm and Y0.25 mm, the error would be:

$$\text{Error} = \sqrt{2 \cdot (25)^2} \cdot \sin 18.435° = 0.039 \text{ mm}.$$

This value is sufficient for rough milling and for some finishing applications. The steps could be further reduced; however, the programming time and the length of the tape would become prohibitive.

Straight-cut positioning controls can be built with open- or closed-loop feedback drives, with programmable accuracies of 0.002 mm (0.0001 inch). These systems are most often built with programmable spindle speed and with various preparatory (G) codes.

EXERCISE

Calculate the increments to be programmed to produce a tool path with less than 0.03 mm error from P1 (11.32, 9.82) to P2 (113.82, 64.32).

3.6 CONTOURING, OR CONTINUOUS PATH CNC SYSTEMS

The contouring system is the state-of-the-art, high-technology, most versatile and intricate of the CNC devices.

It generates a continuously controlled tool path, by interpolating intermediate points or coordinates. By "interpolating" we mean the capability of computing the points of the path.

All CNC contouring systems have the ability to perform linear interpolation. This feature is a computer subroutine, permanently recorded in the CNC computer under a G01 preparatory code. G01, programmed with X-Y or X-Y-Z dimensions, is an instruction to interpolate a tool path, in its shortest distance between two points.

The best way to illustrate the usefulness of this feature is to compare the programming effort difference between contouring and the previously discussed straight-cut system.

In order to program the tool path from 0 to P1 in Fig. 3-6, we would require a single tape block on a contouring system, as shown below:

N100 G01 X120.00 Y60.00 F. . .

This line will produce a tool path to 0.002-mm tolerance.

Using the straight-cut system, the program shown below will require 240 tape blocks and will only yield a tolerance of 0.039 mm.

N1 . . . . . X0.50 Y0.25	from zero to X0.50, Y0.25
N2 . . . . . X0.50 Y0.25	to X1.00, Y0.50
N3 . . . . . X0.50 Y0.25	to X1.50, Y0.75
. . . . .	
. . . . .	
. . . . .	
N240 . . . X0.50 Y0.25	to X120.00, Y60.00

Most contouring systems also have a circular or parabolic interpolation feature. The programming of circular arcs is also done in one tape block; however, in addition to the appropriate preparatory (G) code, additional dimensions must be programmed to identify the relationship between the start point and the arc center.

Most contouring systems have programming capabilities of four or five axes. A large percentage of contouring CNC systems are built with positioning and velocity feedback drives.

3.7   SELECTING THE "RIGHT" SYSTEM

The CNC industry has been built on a solid scientific foundation by highly reputable companies using the best available research. Control systems are highly reliable and are known to maintain their accuracy. However, for a particular, well-defined application, there is only one "best" system for the return on the capital invested.

For drilling, tapping, or boring type of applications, the control type should be an open-loop point-to-point system. These systems are now built to provide positional accuracies of 0.002 mm (0.0001 inch), with high degrees of repeatability.

If a reasonable percentage of the work planned to be machined on the new system requires straight milling, the control should be straight-cut positioning with open- or closed-loop control.

Should the majority of the jobs require contouring of slopes or circular arcs, the system selected should be a contouring one. While the closed-loop drive system offers a higher degree of accuracy, the open-loop concept should not be disregarded. The price differential significantly favors the open-loop concept, and competent handling of the machine will produce the required accuracy.

As a final consideration, if the manufacturer or the dealer cannot guarantee competent service within 24 hours, the system should not be considered for purchase.

4

Machining Forces

The programming of feeds and speeds is the responsibility of the part programmer. While the programming accuracies relate to the dimensional accuracy of the work piece, the programming of feeds and speeds relates to the efficiency of the machining process. As a result, they carry equal importance in CNC programming. As in the previous paragraphs, we shall discuss only the most important formulas. The most important consideration in the selection of speed and feed is the available horsepower of the CNC machine tool. Generally, the power required for drilling, milling, or turning is expressed in terms of the average unit power. The average unit power is the power required to remove 1 cubic inch of metal in 1 minute.

4.1   DRILLING

4.1.1   Cutting Speed (Vc)

See Fig. 4-1 for illustration of drill area.

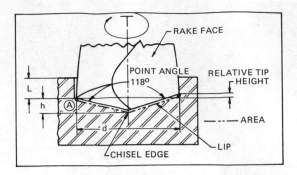

FIG. 4-1. Drill area.

At Point A
In inches:

$$Vc = \frac{d \cdot \pi \cdot n}{12} \text{ feet per minute (fpm)}$$

In metric:

$$Vc = \frac{d \cdot \pi \cdot n}{1000} \text{ m/min.}$$

Where:

Vc = Cutting Velocity
n = Drill rpm
d = Drill Diameter (in inches or mm)

4.1.2 *Rate of Metal Removal (Qd)*

In inches:

$$Qd = A \cdot n \cdot F \qquad \text{in in}^3/\text{min}$$

In metric:

$$Qd = \frac{A \cdot n \cdot F}{1000} \qquad \text{in cm}^3/\text{min}$$

Where:

Qd = Volume of metal removed in in^3/min or cm^3/min
n = Drill rpm
F = Feed in inches or mm per revolution of drill

$$A = \text{Area} = r^2 \cdot \pi = \frac{d^2 \cdot \pi}{4} \text{ in square inches or mm}^2$$

4.1.3 Horsepower At Spindle

$$HPs = k \cdot Qd \text{ at 100\% efficiency}$$

Where:

HPs = Horsepower required for machining

Qd = Volume of material removed in in³/min or cm³/min

k = Material constant, as shown in Fig. 4-2.

MATERIALS	CONSTANTS (k)		
	DRILLING	MILLING	TURNING
MILD STL. 25 RC	1.0	1.0	.9
MILD STL. 25 - 30 RC	1.6	1.8	1.3
HARD STL. 50 RC	1.9	2.1	1.5
SOFT CAST IRON	.8	.7	.5
HARD CAST IRON	.9	1.1	1.0
ALUMINUM	.35	.4	.3
BRASS	.5	.6	.4
BRONZE	.6	.8	.7
STAINLESS 400	1.3	1.3	1.1
STAINLESS 300	1.6	1.8	1.7
TITANIUM	1.0	1.0	1.1
NICKEL ALLOYS	1.6	1.6	1.5

FIG. 4-2. Table of constants.

Since 100% efficiency is a theoretical value, we must in fact alter the above equation in view of the efficiency (E) of the spindle drive. In practice, E may vary from 0.7 ~ 0.85 depending on the condition of the machine tool. The practical formula for calculating the horsepower (hp) will therefore change to:

$$HPm = \frac{k \cdot Qd}{E}$$

where HPm represents the horsepower of the spindle drive motor.

4.1.4 Torque on Spindle Due to Drilling (Ts)

$$Ts = \frac{63030 \cdot hps}{n}$$

NOTE: *Horsepower is the unit of power adopted for engineering use. hp = 33.000 ftlb/min = 550 ft lb/sec. Metric horsepower = 75 kg · m/sec = 542.5 ftlb/sec. While the International System (metric) unit of power is the watt, its applications haven't yet reached the usual engineering tables.*

4.1.5 Machining Time

Machining time is expressed in terms of minutes.

$$T = \frac{L}{F} \quad \text{where } L = \text{Depth of drilling in inches or mm}$$
$$\text{and} \quad F = \text{Feed in ipm or mm/min}$$

4.2 TURNING

4.2.1 Cutting Speed (Vt)

See Fig. 4-3 for illustration of turning.

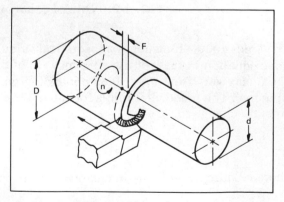

FIG. 4-3. Turning.

$$Vt = \frac{D \cdot \pi \cdot n}{12} \text{ fpm in inch units}$$

or

$$Vt = \frac{D \cdot \pi \cdot n}{1000} \text{ m/min in metric}$$

4.2.2   Rate of Metal Removal (Qt)

$$Qt = T \cdot F \cdot n \cdot c \qquad \text{in in}^3/\text{min or cm}^3/\text{min}$$

Where

Qt = Volume of material removed
F = Feed in ipr or mm/rev
n = Part rpm
T = Depth of cut $\dfrac{D - d}{2}$
C = Circumference $D \cdot \pi$

4.2.3   Horsepower

$$HPm = \frac{k \cdot Q}{E.}; \qquad HPm \text{ is the horsepower of the spindle drive motor.}$$

4.2.4   Torque on Headstock Due to Turning

$$Ts = \frac{63030 \cdot HPs}{n}$$

4.2.5   Surface Roughness

The theoretical surface roughness (Sr) can be calculated by the formula below. However, this should only be used as a programming guideline. Actual surface finish depends on a number of other factors, such as sharpness of tool, coolant, tool part rigidity, etc. Therefore, the calculated result may vary substantially if any or all of the above factors are not ideal.

$$Sr = \frac{F}{8000 \cdot r} \text{ micro inches, or}$$

$$Sr = \frac{F}{8 \cdot r} \text{ micrometers, where}$$

F = Feed rate in ipr or mm/rev
r = Tool nose radius in mm or inches

4.2.6 Acceleration and Deceleration Distance for Thread Turning

The programmer must allow a distance A_1 for the tool to accelerate to a constant (programmed) feed rate and a distance A_2 for deceleration. See Fig. 4-4. Since the tool point velocity (feed rate) will change from zero to the programmed rate in the A_1 interval, the thread cut on this distance will be imperfect. The same condition applies to the A_2 distance.

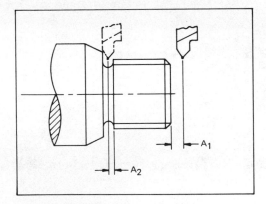

FIG. 4-4. Tool point acceleration-deceleration.

The formulas below can be used as a guideline if specific formulas are not provided by the manufacturer of your CNC system.

The value of the time constant t_1 of the system, provided by the manufacturer, is a function of the specific system dynamics. Factors such as the weight of the tool slide or turret, coefficient of friction, size and torque of the servodrive, servogain, etc., influence the specific CNC time constant.

The Fanuc 4NE system, used in our example, has a time constant $t_1 = 0.12$. Multiplying this by the linear velocity of the tool point, $A_2 =$

$t_1 \cdot V_L$, allows us to calculate the distance required to decelerate the tool from the programmed feed (velocity) to zero.

EXAMPLE

Calculate the deceleration distance required to turn a 1½"–8 tpi thread at 350 rpm.

Solution:

$$8 \text{ } tpi = 0.125 \text{ } ipr,$$

The linear velocity of the tool point will be 0.125 inch for each revolution.

The linear velocity of the tool point at 350 rpm:

$$V_L = 0.125 \cdot 350 = 43.75 \text{ } ipm = 0.729 \text{ } ips$$

$$A_2 = t_1 \cdot V_L = 0.12 \cdot 0.729 = 0.0875 \text{ inches}$$

which is the deceleration distance.

The acceleration distance calculation requires a value for the acceleration time constant, t_2. This is calculated as follows:

$$t_2 = t - t_1 + t_1 \cdot EXP\left(\frac{-t}{t_1}\right)$$

Where

$$EXP\left(\frac{-t}{t_1}\right) = 0.05$$

EXP is the abbreviated version of "e at the power of . . . ," where $e = 2 \cdot 7182818$.

Using natural logarithms on both sides of the equation, we obtain

$$\frac{-t}{t_1} = \ln 0.05 = -2.99573$$

$$-t = t_1 \cdot (-2.99573)$$

or

$$t = 0.12 \cdot 2.99573 = 0.3595$$

Substituting this value into the above equation we obtain:

$$t_2 = 0.3595 - 0.12 + 0.3594 \cdot 0.05 = 0.2455$$

The acceleration distance can now be calculated from the formula:

$$A_1 = t_2 \cdot V_L$$

EXAMPLE

Calculate the acceleration distance required to turn a 1½"–8 tpi threads at 350 rpm.

Solution:

As in the preceding example,

$$8 \; tpi = 0.125 \; ipr$$
$$V_L = 0.125 \cdot 350 = 43.75 \; ipm = 0.729 \; ips$$
$$\text{and} \quad A_1 = t_2 \cdot V_L = 0.2455 \cdot 0.729 = 0.1789 \; \text{inches}$$

This is the acceleration distance.

The A_1 distance varies between 2 to 4.4 times the A_2 for CNC systems.
or

$$A_1 = 2 \sim 4.4 \cdot A_2$$

4.3   MILLING

4.3.1   Cutting Speed (Vm)

$$Vm = \frac{D \cdot \pi \cdot n}{12} \, fpm, \text{ where the diameter } D \text{ is expressed in inches.}$$

and

$$Vm = \frac{D \cdot \pi \cdot n}{1000} \, m/min, \text{ where } D \text{ is in mm.}$$

4.3.2   Rate of Metal Removal (Qm)

$$Q_m = W \cdot T \cdot F \; in^3/min \text{ or}$$
$$Q_m = \frac{W \cdot T \cdot F}{1000} \; cm^3/min \text{(metric)}$$

Where

Q_m = Volume of metal removed in in³/min or cm³/min, as applicable.
W = Width of cut in inches or mm
T = Depth of cut in inches or mm
F = Programmed feed in ipm or mm/min.

4.3.3   Horsepower

$$HPm = \frac{k \cdot Q}{E}$$

4.3.4   Torque on Spindle

Torque on the spindle can be calculated as

$$Ts = \frac{63030 \cdot hps}{n}$$

The calculations and formulas presented in this chapter are for ideal machining conditions. The reader must adjust the calculated values if the rigidity of the setup is not as good as desired or if the tool deflection, due to the cutting forces, create undesirable (destructive) vibrations.

5

Cutter Centerline Programming

Before getting involved in the actual programming, we shall briefly review some of the very basic codes and principles.

ABSOLUTE PROGRAMMING: This is a mode of programming in which an origin has to be selected for each axis prior to starting the program. Once this "part program origin" has been selected, all motions have to be stated with respect to this origin. What this means is that all motions defined in the program are in reality "locations" or "addresses" of the particular point defined in relation to the origin. Usually this is the mode of programming selected by experienced programmers, as the programmed values match fairly closely the values defined on the engineering drawing. The drawing's datum point is a direct equivalent of the program origin.

In milling programming, the absolute mode is established by the use of the preparatory code G90. In turning, the use of the tape words X and Z means that the cross direction and the longitudinal direction, are programmed in absolute, without requiring a particular preparatory code.

INCREMENTAL PROGRAMMING: When programming in incremental, an origin need not necessarily have been selected. All motions are stated from the immediate last position of the tool. This means that all incremental motions are "displacements" from a given position. The immediate short-range advantage of incremental programming is that the programmed motion matches directly the actual motion of the tool, as both take place from the last previous position of the cutter. Another advantage is that the sign of the motion is also directly related to the tool motion. In incremental programming, a positive dimension will cause a tool motion in the appropriate direction, and a negative one will take place in the opposite sense. In absolute programming, the sign depended on the quadrant the tool moved in, not in the direction of travel.

In milling programming, incremental is programmed using the preparatory code G91. In turning, incremental may be programmed the same way, or using the letters U and W to replace X and Z, in which case the preparatory code G91 is not required.

PROGRAMMING THE ORIGIN: Establishing the origin of the part program for subsequent use in absolute programming is known under several terms such as register preset, work origin setting, program zero point, or position absolute coordinates setting. This proliferation of terminologies exists because NC technology is fairly new, and yet it has expanded too rapidly to allow standardization to catch up with it.

A program origin and a coordinate system are a must. The program origin preparatory code, G92 in milling and G50 in turning, will not cause any motion. The code will tell the system where the tool is located at that given time in relation with the program zero. In reality, the machine knows where the tool is and does not know where the program origin is. Therefore, when we tell the machine where the tool is in relation with the work zero, we are really telling it where the work origin is in relation to the known position of the tool. From there on, absolute programming will give addresses from this datum point.

RAPID: The G00 (G zero zero) will result in rapid positioning of the programmed machine slides to the required location. It is used for rapid approaching of the part, or for rapid moves between holes in drilling or boring applications. The tool or the table will always travel at the highest machine speed. It should be borne in mind, however, that regardless of the actual element of the machine which provides the move, in our program it is the tool that moves in relation to the part.

In most controls (there are some recent exceptions), the rapid motion will follow the "point-to-point" pattern, i.e., all programmed axes will start simultaneously, and as one is achieved, its motor will stop while the others will continue as required.

A safety reminder may be in order for people who sometimes lean on machine components. It is not unusual for a CNC machine to move in Rapid at 10 ips (approximately 250 mm/s).

LINEAR INTERPOLATION: This feature is programmed using the preparatory code G01 (G zero one), in conjunction with the appropriate dimensional tape words, as well as a programmed feed. The corresponding machine axes, with their own variable speed-controlled drive systems independent from each other, will produce the required straight-line motions by driving the slides at different speeds.

CIRCULAR INTERPOLATION: Two codes are used in programming a circular arc, G02 (clockwise) and G03 (counterclockwise). The CNC system has the capability to establish and maintain the relative positions and velocities of two machine slides, on a constantly changing basis, but starting and stopping at the same time.

At the start of circular interpolation, the control knows the position of the cutter. It must be told the desired position at the end of the programmed arc, in either incremental or absolute coordinates. The control also has to know the location of the center of arc and the value of the arc radius. This may be achieved using I and J or K tape words in older controls, or R (for radius) in the newer ones. Subsequent programming examples in this chapter will illustrate this particular technique.

This chapter will discuss, in some detail, cutter centerline programming. Its importance is underscored by the fact that older systems can only be programmed in cutter centerline. Newer systems, with advanced cutter compensation features, are much easier to comprehend if cutter centerline programming is known and understood. In addition, in some special cases, of very intricate parts, the use of the compensation features may present a complication, and the programmer can solve the problem and get the job done by using an "old-fashioned" programming method on a state-of-the-art CNC system.

The program will guide the cutter around the part contour. The cutter will have to follow the path at a set distance away from the part, at every point, corresponding to the cutter radius.

Parts with complex geometry will require a certain level of calculations, mainly trigonometry, so a review of the math chapter may be necessary.

5.1　CALCULATING CUTTER CENTERLINE DISTANCES

To write a part program for the outside contour milling of the job illustrated in Fig. 5-1, we shall first have to select a cutter (end mill) diameter.

　　In cutter centerline mode, the cutter center is programmed, yet the actual cutting is performed by the edge of the cutter. In order to obtain the required shape of the part, the cutter must be correctly placed at each point (1, 2, 3 . . . . to 8 and back to 9) and of course in between, as shown in Fig. 5-1.

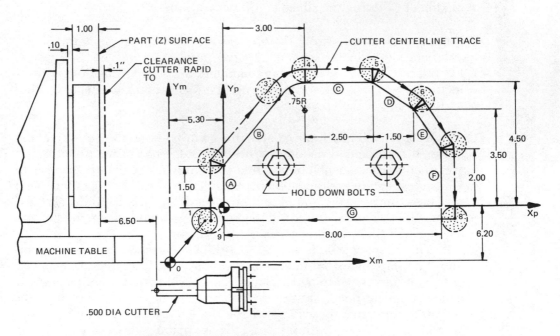

FIG. 5-1. Cutter centerline programming.

　　This placement is that of a circle (i.e. the cutter) tangent to a line or circle representing the part contour.

　　The line connecting the cutter center to the cutting edge at the part (cutter radius) will be perpendicular to the part contour being cut.

　　This requirement of perpendicularity at a known location, combined with the knowledge of the cutter radius (half the cutter diameter) will enable us to calculate the cutter center location at each point by solving appropriate right triangles, as shown later in the chapter. The program drives the machine spindle where the center coincides with the cutter center.

It should be borne in mind that regardless whether the cutting action is achieved by moving the cutter or the part, in programming it is the cutter that is assumed to be mobile and the part that is considered stationary. Always *think tool!*

Having calculated these dimensions, we must then convert them into programmable distances. The actual values of these distances will be different depending on whether we use incremental or absolute programming. The calculation of the values may take up more time than the actual programming, and the reader may be tempted to skip straight to cutter compensation or computer-assisted programming. However, neither can be properly understood and used without a strong foundation in cutter centerline programming.

In the following, we shall discuss the necessary steps required to calculate the center point locations of a 0.500-inch-diameter cutter, in "incremental" programming mode. We will also assume that the part has been bandsawed to size, leaving approximately 0.150 inch for finish machining, and that the material is SAE 1030.

The programming methods illustrated below can be used for any similarly shaped part so long as the machine codes have been double-checked for the actual system used.

START UP

N0010 G20 G40 G80 G91	Inch programming (G20 or G70), cancel cutter diameter compensation, cancel canned cycles, incremental programming. The "cancels" are for safety.

As mentioned at the beginning of this chapter, inexperienced programmers should start programming in incremental (G91). The incremental motion is a "displacement" of the tool from its present location, as opposed to absolute, which takes the tool to an address or a location defined in relation to an origin. Incremental provides a direct correlation between programmed motion and tool motion.

N0020 G92 X0 Y0 Z0	Set control display to zero, to correspond to the machine "zero" position.
N0030 S800 M03	Spindle speed 800, spindle start clockwise
N0040 G00 X4.75 Y5.95	"Rapid" to part in X and Y

N0050 Z-6.4	Rapid to part in Z, leaving 0.1 inch clearance
N0060 G01 Z-1.15 F3.0	Feed 0.05 inch past part at 3 ipm
N0070 X0.3 M08	Feed to point 1, coolant on

5.1.1 Machine Part Surface "A" Between Point 1 and Point 2

Since the necessary "Y" motion is not readily available on the part drawing, we shall have to calculate it from the given dimensions. There is no "X" motion from point 1 to point 2. However, the "Y" motion will be longer than the 1.5-inch length of surface "A" by an amount $\Delta Y2$ (see Fig. 5-2).

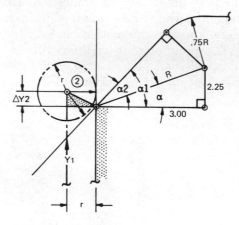

FIG. 5-2. Cutter center at point 2.

This calculation is necessary in order to position the cutter radius "r" perpendicular to the upcoming surface "B."

$$\Delta Y2 = r \cdot \tan\left(45° - \frac{\alpha 1}{2}\right) \qquad \ldots \ldots \ldots 5\text{-}1$$

$$\tan \alpha = \frac{2.25}{3.00} = 0.75$$

$$\alpha = 36.8699°$$

$$R = \sqrt{3.0^2 + 2.25^2} = 3.750$$

$$\sin \alpha 2 = \frac{0.75}{3.75} = 0.2$$

$$\alpha 2 = 11.5369°$$

$$\alpha 1 = \alpha + \alpha 2 = 36.8699 + 11.5369 = 48.4068°$$

Substituting $\alpha 1$ into equation 5-1, we obtain

$$\Delta Y2 = r \cdot \tan \left(45 - \frac{48.4068}{2} \right)$$

$$= 0.25 \cdot \tan 20.7966 = 0.0949 \text{ inch}$$

Having calculated the $\Delta Y2$ value, we can now derive the Y2 value, to be programmed as the "Y" motion from point 1 to point 2.

$$Y2 = r + 1.5 + \Delta Y2 = 0.25 + 1.5 + 0.0949 = 1.8449 \text{ inches}$$

and the next program line will be

N0080 Y1.8449   Move cutter, in feed, to point 2

5.1.2   Machine Part Surface "B" to Point 3 and Arc to Point 4

We must now calculate the dimensions X3 and Y3 from Fig. 5-3, prior to programming the tool motion from point 2 to point 3. These dimensions again cannot be readily taken off the drawing, as is the case in most programming applications. Calculations similar to the ones above are required for arc to straight-line translations.

From the previous calculations we retain:

$$\alpha 1 = 48.4068°$$
$$\Delta Y2 = 0.0949 \text{ inch}$$
$$r = 0.25 \text{ inch (the cutter radius) and}$$
$$R = 0.75 \text{ inch (the part radius)}$$
$$\Delta Y3 = r \cdot \quad \text{s } \alpha 1 = 0.25 \cdot \cos 48.4068° = 0.1659 \text{ inch}$$
$$\Delta J = R \cdot \cos \alpha 1 = 0.75 \cdot \cos 48.4068 = 0.4978 \text{ inch}$$
$$Y3 = 2.25 + \Delta J + \Delta Y3 - \Delta Y2$$
$$= 2.25 + 0.4978 + 0.1659 - 0.0949 = 2.8188 \text{ inches}$$

This will be the Y-motion from point 2 to point 3.

$$\Delta Y = r \cdot \cos \alpha 1 \qquad \dots \dots \dots \dots \ 5\text{-}2$$
$$\Delta X = r \cdot \sin \alpha 1 \qquad \dots \dots \dots \dots \ 5\text{-}3$$
$$\Delta j = R \cdot \cos \alpha 1 \qquad \dots \dots \dots \dots \ 5\text{-}4$$
and
$$\Delta i = R \cdot \sin \alpha 1 \qquad \dots \dots \dots \dots \ 5\text{-}5$$

$$j = \Delta j + \Delta Y3$$
and
$$i = \Delta i + \Delta X3$$
$$\Delta X3 = r \cdot \sin \alpha1 = 0.25 \cdot \sin 48.6068° = 0.1875 \text{ inch}$$
$$\Delta j = R \cdot \sin \alpha1 = 0.75 \cdot \sin 48.6068° = 0.5625 \text{ inch}$$
$$X3 = 3.0 + r - (\Delta X3 + \Delta i)$$
$$= 3.0 + 0.25 - (0.1875 + 0.5626)$$
$$= 3.25 - 0.7501 = 2.4999 \text{ inches}$$

This will be the X-displacement from point 2 to point 3, and the next program line will be

N0090 X2.4999 Y 2.8188   To point 3.

To machine the 0.750-inch radius arc, we have to program the X- and Y-displacements from the start of the arc to the end point of the arc, as well as the values "i" and "j," required to position the center of arc in relation to the start point of the arc. Point 3 is not on either of the primary axes (X or Y) of the system of coordinates. It may be worth pointing out at this time that as NC systems changed, they became progressively easier to program. Unlike many of the old machines still in use, the modern CNCs have no constraints of primary axes or i and j unit vectors for circular interpolation, trailing zeros, etc. Many of the newer systems accept the "old" formats as well, but the contrary is not true. Circular interpolation is normally programmed using direct radius programming, which is much easier. This program, however, will be done the hard way.

The calculations for the next tool motion are illustrated in Fig. 5-3.

$$X4 = \Delta i + \Delta X3 = 0.5626 + 0.1875 = 0.7501 \text{ inch}$$

This will be the X-displacement from point 3 to point 4.

$$Y4 = R + r - (\Delta j + \Delta Y3) = 0.75 + 0.25 - (0.4978 + 0.1659)$$
$$Y4 = 0.3363 \text{ inch}$$

This will be the Y-displacement from point 3 to point 4

$$i = X4 = \Delta i + \Delta X3 = 0.7501 \text{ inch}$$

The dimension "i" is called a "unit vector," and it is measured from the start point 3 of the arc, where the cutter is located, to the center of the arc, to be measured parallel to the X-axis.

$$j = \Delta j + \Delta Y3 = 0.4978 + 0.1659 = 0.6637 \text{ inch}$$

The dimension "j," also called a "unit vector," is measured parallel to the Y-axis from the start point of the cutter to the center of the arc. For the reader who has not at this point developed an understanding of the cir-

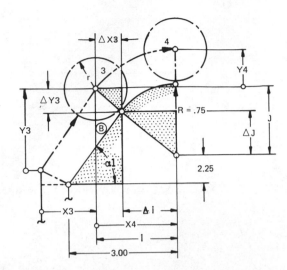

FIG. 5-3. Cutter centerline geometry at straight-line and circular arc intersection.

cular interpolation process, the following explanation might be of some help. The cutter is in position prior to the start of the arc, and the control knows its position, since this was the end point of the previous travel motion. From the start point of the arc, the control will travel an imaginary path along "j," then along "i." At this point the computer has found the center of the arc. Using the values i and j as sides of a right-angle triangle, the control will compute the hypotenuse of the triangle, which is the radius of the arc. Knowing the center and the radius, the control will now trace an imaginary circular path, along which the cutter will travel a real distance defined by the programmed dimensions X and Y.

The circular interpolation motion block can be written therefore as follows:

N0100 G02 X0.7501 Y0.3363 I0.7501 J-0.6637

The unit vectors can be spelled in either lowercase or capital letters.

5.1.3 Machine Part Surface "C" from Point 4 to Point 5

The calculations for machining the part surface "C" from point 4 to point 5 will be identical to the ones corresponding to point 2.

The offset will be in the X-direction by the amount $\Delta X5$.

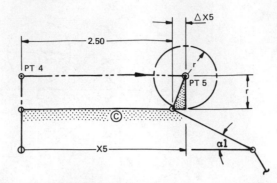

FIG. 5-4. Cutter centerline geometry at point 5.

$$\Delta Y5 = r$$

$$\Delta X5 = r \tan \frac{\alpha 1}{2} \quad \ldots \ldots \ldots \ldots \ldots \quad 5\text{-}6$$

From the part dimensions,

$$\text{Tan } \alpha 1 = \frac{1.00}{1.50} = 0.66666, \text{ or } \alpha 1 = 33.59°$$

Using the triangle shown in Fig. 5-4 and formula 5-6,

$$\Delta X5 = r \cdot \tan \frac{\alpha 1}{2} = 0.25 \cdot \tan \frac{33.59}{2} = 0.0754 \text{ inch}$$

$$X5 = 2.50 + \Delta X5 = 2.50 + 0.0754 = 2.5754 \text{ inch}$$

This will be the X-motion from point 4 to point 5, and the corresponding program line will read:

N0110 G01 X2.5754

5.1.4 Machine Surface "D" from Point 5 to Point 6

To machine the surface "D" from point 5 to point 6, the cutter motion will require offsets in both X- and Y-directions, since neither surface "D" nor "E" is parallel with the primary axes of the system. The angles $\alpha 1$ and β will have to be measured from a theoretical line, parallel with the X-axis, as shown in Fig. 5-5. From Fig. 5-4, $\alpha 1 = 33.59°$

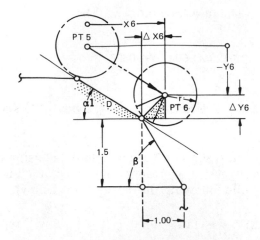

FIG. 5-5. Cutter centerline geometry at point 6.

$$\Delta Y6 = \frac{\cos\dfrac{\alpha1 + \beta}{2}}{\cos\dfrac{\alpha1 - \beta}{2}} \cdot r \quad \ldots \ldots \quad 5\text{-}7$$

$$\Delta X6 = \frac{\sin\dfrac{\alpha1 + \beta}{2}}{\cos\dfrac{\alpha1 - \beta}{2}} \cdot r \quad \ldots \ldots \quad 5\text{-}8$$

$$\text{Tan } \beta = \frac{1.5}{1} = 1.5 \quad \text{or} \quad \beta = 56.3099°$$

$$\Delta Y6 = \frac{\cos\dfrac{\alpha1 + \beta}{2}}{\cos\dfrac{\alpha1 - \beta}{2}} \cdot r = \frac{\cos\dfrac{33.59 + 56.31}{2}}{\cos\dfrac{33.59 - 56.31}{2}} \cdot 0.25$$

$$= \frac{\cos 44.95}{\cos(-11.36)} \cdot 0.25 = \frac{0.707724}{0.980409} \cdot 0.25 = 0.1804 \text{ inch}$$

$$\Delta X6 = \frac{\sin\dfrac{\alpha1 + \beta}{2}}{\cos\dfrac{\alpha1 - \beta}{2}} \cdot r = \frac{\sin 44.95°}{\cos(-11.36)} \cdot r = \frac{0.706489}{0.980409} \cdot 0.25$$

$$= 0.1801 \text{ inch}$$

X6 = 1.5 + ΔX6 − ΔX5 = 1.5 + 0.1801 − 0.0754 = 1.6047 inch
and
Y6 = 1.00 + r − Y6 = 1.00 + 0.25 − 0.1804 = 1.0696 inch

The next tape block will be:

N0120 X1.6047 Y-1.0696

5.1.5   Machine Surface "E" to Point 7

The offset at point 7 can be calculated using equation 5-1. As an exercise, it is suggested the reader draw an enlarged figure of the geometry at point 7, while following the calculations below:

$$\Delta Y7 = r \cdot \tan \left(45° - \frac{\alpha 1}{2} \right) \qquad \beta \text{ will replace } \alpha 1$$

$$\Delta Y7 = 0.25 \cdot \tan \left(45° - \frac{56.31}{2} \right) = 0.25 \cdot \tan 16.845° = 0.0757 \text{ inch}$$

$$X7 = 1.00 + r - \Delta X6 = 1.00 + 0.25 - 0.1801 = 1.0699 \text{ inches}$$
$$Y7 = 1.5 + \Delta Y6 - \Delta Y7 = 1.50 + 0.1804 - 0.0757 = 1.6047 \text{ inches}$$

The program line will read:

N0130 X1.0699 Y-1.6047

5.1.6   Machine Surface "F" from Point 7 to Point 8

$$X8 = 0$$
$$Y8 = 2.00 + r + \Delta Y7 = 2.00 + 0.25 + 0.0757 = 2.3257 \text{ inches}$$

The program line is:

N0140 Y-2.3257

5.1.7   Machine Surface "G" from Point 8 to Point 9

$$Y9 = 0$$
$$X9 = 8.0 + r + 0.05 = 8.0 + 0.25 + 0.05 = 8.3 \text{ inches}$$

The part would be machine finished without the additional 0.05-inch-increment. It is, however, a desirable practice to move the cutter past the finished surface by 0.01 to 0.1 inch.

The line of program corresponding to surface "G" is

N0150 X-8.3

5.1.8   *Return Tool to the Machine Zero ("Home") Position*

N0160 Z1.15 F0.4 M09   Clear part, stop coolant
N0170 G28 Z0.1   Return Z "home"
N0180 G28 X-0.1 Y-0.1   Return X and Y "home"

The few basic calculations discussed so far will not cover more complex part surface intersections, such as the tangency of two circular arcs or an arc tangent to a line. These will come up later, when more complex part geometries will be studied under CNC turning.

5.2   TOOL NOSE RADIUS CENTERLINE CALCULATIONS FOR CNC TURNING

Tool nose radius (TNR) calculations are also known as "equidistance programming." When CNC turning is first discussed, TNR is usually overlooked. However, once the basics have been established, we must remember that every turning tool tip has a radius, small as it may be. See Fig. 5-6. This radius will have to be taken into consideration when the turned part surfaces are not parallel with the primary axes of the machine (X and Z in the case of the turning center). The error will be the greatest when turning a surface at 45° from the primary axes.

The actual inaccuracy or error, E, can easily be calculated if we know the radius r of the tool tip. Assuming a value of $r = 0.015$ inch, for example,

$$\text{the error} \quad E = H - r = \frac{r}{\sin 45°} - r = \frac{0.015}{0.7071} - 0.015 = 0.0062 \text{ inch}$$

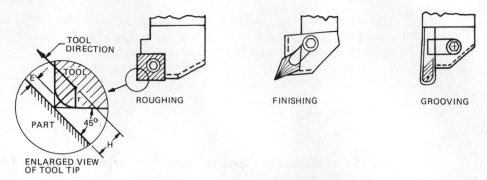

TOOL DIRECTION

TOOL

E

PART 45°

H

ENLARGED VIEW OF TOOL TIP

ROUGHING

FINISHING

GROOVING

FIG. 5-6. Typical turning tool tips.

The value of $H = \dfrac{r}{\sin 45°}$ results from Fig. 5-6.

The tapered part surface diameter at a 45° angle would therefore be larger by $2 \cdot 0.0062 = 0.0124$ inch. This deviation would not be acceptable from a quality control point of view.

It can be seen at this stage that this inaccuracy problem is more complex in the case of turning than it is for milling, because of its variability. The error disappears completely when turning surfaces parallel to either the X- or Z-axis of the turning center. For tapered surfaces, the error becomes proportional to the part surface angle, the deviation increasing from zero part surface angle to its maximum value at 45°. In addition, the error will further increase as the TNR becomes larger.

To eliminate this tool tip error in NC turning systems without programmable TNR compensation, we must, prior to programming, calculate the path of the tool tip radius center. The method of calculation is similar—in fact in many cases identical—to the one discussed in section 5.1.

To illustrate the different steps to follow, the equidistance calculations will be carried out in parallel with the programming for the part shown in Fig. 5-7.

In the following, we have used the *right-hand coordinate system*, common to most CNC turning centers. In the right hand system, $+Z$ points away from the chuck, $+X$ points away from the operator, G02 is circular interpolation clockwise, G03 is circular interpolation counterclockwise, G41 is tool nose radius compensation left and G42 is tool nose radius compensation right. The standard tool nose vector chart is as illustrated in Fig. 7-6.

Some turning centers use the *left-hand coordinate* system. The essential difference is that $+X$ now points towards the operator. This change reverses G02 which becomes counterclockwise, G03 which becomes clockwise, G41 which becomes right and G42 which becomes left.

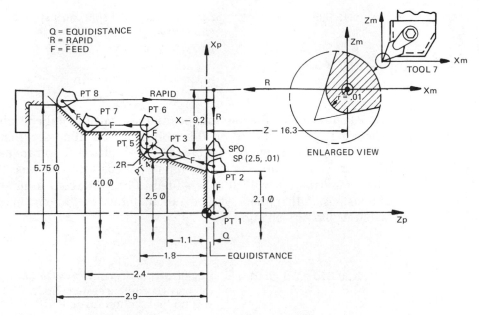

FIG. 5-7. Sample part with equidistance tool trace.

The changes in the chart from Fig. 7-6 are as follow: point 2 changes place with 3, 1 with 4, and 6 with 8.

As in the case of milling, it will be assumed that the part has been previously machined, in our case rough turned. The calculations will be performed and the program written for the turning of the finished contour as dimensioned. Equidistance calculations are required from point 1 through to point 8. As we proceed with the calculations and the program, the geometry will be drawn in sufficiently enlarged scale so that the necessary trigonometric labeling can be clearly analyzed for each step along the path.

5.2.1   Start Up

It will be assumed that the tool slides have been brought to "home" position.

N01 G50 X0 Z0	"Zero" control to machine home
N05 M41	Select middle speed range
N10 G98 G20 T0700	Feed in ipm, inch programming, turret tool position no. 7

N15 S950 M03	Spindle speed 950 rpm, turn spindle on clockwise.
N20 G00 Z-16.29	Rapid to point 0 in Z
N25 X-9.2	Rapid to point 0 in X (diameter programming)

Having moved the tool into position for cutting in the machine coordinate system, we should, at this point, transfer the coordinate system onto the part as shown in Fig. 5-8. This will allow us to program all the necessary slide motions in terms of the actual part dimensions. The calculated equidistances will thus be added to or subtracted from actual part dimensions, rather than worked out in terms of dimensions from the machine "home" position.

| N30 G50 X2.5 Z0.01 | Established new coordinates in terms of the part coordinate system XpZp. There was no tool motion involved. |
| N35 G01 X-0.01 F2.0 | Turn face |

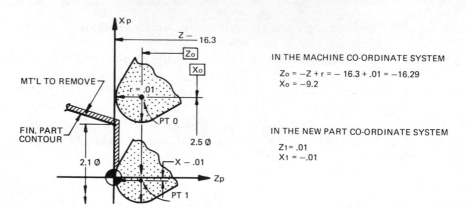

IN THE MACHINE CO-ORDINATE SYSTEM

$Z_o = -Z + r = -16.3 + .01 = -16.29$
$X_o = -9.2$

IN THE NEW PART CO-ORDINATE SYSTEM

$Z_1 = .01$
$X_1 = -.01$

FIG. 5-8. Equidistance at points 0 and 1.

5.2.2   Move Tool to Point 2

The next surface being tapered, the X-position of the tool must be calculated. As shown in Fig. 5-9, no calculations are required for Z, whose position remains tangent to the part face.

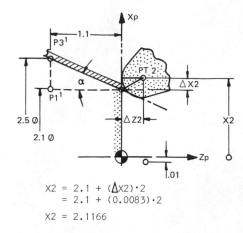

$$X2 = 2.1 + (\Delta X2) \cdot 2$$
$$= 2.1 + (0.0083) \cdot 2$$
$$X2 = 2.1166$$

FIG. 5-9. Equidistance at point 2.

The following important geometric condition arises at point 2: The TNR must be tangent to both current and upcoming part surfaces. From equation 5-1,

$$\Delta X2 = r \cdot \tan\left(45° - \frac{\alpha}{2}\right) = 0.01 \cdot \tan\left(45° - \frac{10.305}{2}\right) = 0.0083 \text{ inch}$$

where α was calculated as $\alpha = \text{InvTan} \dfrac{(2.5\text{-}2.1)/2}{1.1} = \text{InvTan } 0.18181 = 10.305°$ and the next block of tape will be:

N40 X2.1166 F15.0   Move to point 2 (the value of the X-coordinate was calculated in Fig. 5-9)

NOTE: *Your calculator may be labeled Invtan, Arctan, Tan⁻¹ or it may use a function key to calculate an angle when the trigonometric function is known.*

5.2.3 Turn Tapered Surface from Point 2 to Point 3

To maintain the equidistance condition at point 3, the Z3 motion will have to be adjusted by a calculated $\Delta Z3$.

As the surface between points 3 and 4 is parallel to the Z-axis, the X-adjustment will be made by adding the value of the radius to the corresponding dimension. Since X is programmed as a diameter, the adjustment will have to be $2r$.

The geometric conditions at point 3 are as follows:

- The tool tip radius r is perpendicular to the line between points 2 and 3.
- The tool tip radius is tangent at both surfaces, current and next.
- The tool motion between points 2 and 3 is not parallel to either of the primary axes. See Fig. 5-10.

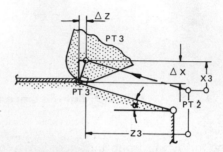

FIG. 5-10. Equidistance at point 3.

$$\Delta Z = r \cdot \tan \frac{\alpha}{2} = r \cdot \tan \frac{10.305}{2} = 0.0009 \text{ inch}$$

$$X3 = 2.5 + 2r = 2.5 + 0.020 = 2.520 \text{ inches}$$

This is the X-motion to point 3.

$$Z3 = 1.1 - \Delta Z = 1.1 - 0.0009 = 1.0991 \text{ inches}$$

This is the Z-motion to point 3.

The next program block will therefore read:

N45 X2.52 Z-1.0991 F7.5

The inquisitive reader has noticed by now that the X- and Z-dimensions have been programmed all along in absolute mode. The calculated values have been added to or subtracted from the actual part dimensions which are defined from two datum lines (axes) intersecting at the part origin (see Fig. 5-7).

5.2.4 Turn 2.5-Inch Diameter to Point 4 and 0.2-Inch Radius to Point 5

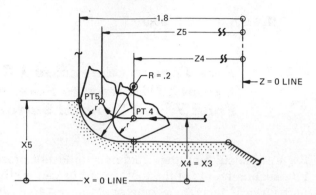

FIG. 5-11. Equidistance at points 4 and 5.

$X4 = X3 = 2.52$ inches; there are no changes in the X-direction
$Z4 = 1.8 - R = 1.8 - 0.2 = 1.6$ inches

The next program block will read:

N50 X2.5 Z-1.6

The program is in absolute mode, and the same motion can be programmed as follows:

N50 Z-1.6
$X5 = 2.5 + 2 \cdot R = 2.5 + 0.4;$ Note that usually "R" refers to the part while "r" is the TNR

$X5 = 2.9$ inches
$Z5 = 1.8 - r = 1.8 - 0.01 = 1.79$ inches
$R = 0.2$

The next block reflects the ability of an advanced control to perform circular interpolation (clockwise) in "radius" programming. We could have, of course, used I and J unit vectors, as in the milling example, with the same practical results. However, the additional calculations may increase the chances of errors, and the additional "characters" and "words" will lengthen the program tape. See Fig. 5-11.

N55 G02 X2.9 Z-1.79 R0.2

5.2.5 Turn Face to Dimension 1.8 Inches and 4.0-Inch Diameter to Point 6, Diameter to Point 7, and Tapered Surface to Point 8

By now, calculating these points should not present any difficulty. The intersecting points of the tool tip radius can easily be projected from the part surface lines. The geometry of the part-tool relationship is illustrated in Fig. 5-12.

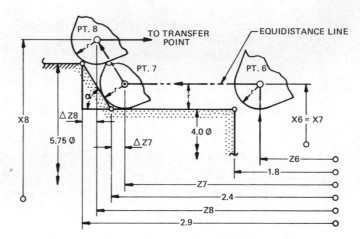

FIG. 5-12. Equidistance at points 6, 7, and 8.

$$X6 = 4.0 + 2r = 4.0 + 0.02 = 4.02 \text{ inches}$$
$$Z6 = 1.8 - r = 1.8 - 0.01 = 1.79 \text{ inches}$$

The next block contains a G01 in order to return to linear interpolation from the preceding circular interpolation. Overlooking this G01 is one of the most frequently made mistakes.

The program block will look as follows:

N60 X4.02 Z-1.79

There is no "Z-address" change, i.e., no Z-motion is required from point 5 to point 6. In absolute programming, the axis that does not contain a displacement can be left out, thus shortening the program. Accordingly, block 60 can be rewritten as shown below:

N60 G01 X4.02

Continuing the calculations,

$$\text{Tan } \alpha = \frac{\dfrac{5.75 - 4.0}{2}}{2.9 - 2.4} = \frac{0.875}{0.5} = 1.75, \text{ or } \alpha = 60.255°$$

$$\Delta Z7 = r \cdot \tan \frac{\alpha}{2} = 0.01 \cdot \tan 30.1275° = 0.0058 \text{ inch}$$

Accordingly,

Z7 = 2.4 − ΔZ7 = 2.4 − 0.0058 = 2.3942 inches
X7 = X6, and the motion to point 7 can be programmed as follows:

N65 Z-2.3942

The next block will be a linear interpolation in both axes. Therefore, the location of point 8 will also have to be calculated. If you observe the tool positions at points 7 and 8, you will note that the angle "α" is shared.

ΔZ8 = ΔZ7, already calculated to be 0.0058 inch

Hence ΔZ8 = 0.0058 inch

X8 = 5.75 + 2r = 5.75 + 0.02 = 5.77 inch

and finally

Z8 = 2.9 − ΔZ8 = 2.9 − 0.0058 = 2.8942 inches

The tape block for the motion to point 8 will read:

N70 X5.77 Z-2.8942

To conclude the program, the motions from the part system will be retransferred to the machine coordinate system as follows:

N75 G00 Z0.01	Motion to Z-transfer point
N80 X2.5 M09	Motion to X-transfer point, coolant off
N85 G50 X-9.2 Z-16.29	No motion, just transfer of the datum point.
N90 G28 X0 Z0 M05	Return "home," spindle off
N95 M30	End of program

6

Tool Offsets

Most CNC machines have a set location, sometimes two, known as the "home" position, the "machine origin," or the "machine absolute origin." These expressions are used interchangeably. For the purpose of this book, we shall use the term "machine origin." The location is physically established by the manufacturer usually by a dual-limit switch system. The first one sets the motion deceleration, and the second one stops it.

The machine origin is used by manufacturers to synchronize the machine with the control, and to establish a start point for measuring the length of travel in the various axes. Some recent systems cannot even be started at all until this synchronization, known as the "zeroing," or "referencing" has been performed. The zeroing is carried out by returning the machine to "machine origin" and initializing the control manually or through programmed statements, such as G92 X0 Y0 Z0, or others, as applicable. By machine we understand the table, column, saddle, gantry, tool post, or whatever moving parts are involved. Initializing the control means setting the X-, Y-, and Z-displays (counters) to zero.

In addition to the machine origin, we have the following additional potential origins:

- the fixture origin
- the part origin
- the program origin

All these origins could be the same, or different, or grouped in some way. To show the flexibility of the systems, we could have more than one fixture or more than one part on the machine. In the latter case, the parts could be the same or different.

Over the years, as NC machines went through different phases, the term "zero offset" was used as an improvement over the "fixed zero," and sometimes called "zero shift," "full zero shift," or "floating zero." The objective was, of course, the ability to position the part at some advantageous location on the machine table and "offset" the "zero" to the right place.

The word "offset" has also been used in offset banks or offset registers. The terms "bank" and "register" are interchangeable, and they mean storage areas in the electronic control. These registers can be used for a number of purposes, such as producing small numerical changes to compensate for minute setup adjustments, dimensional inaccuracies, or tool wear. But they can also be used to store specific corrections related to a particular tool, tool radius, tool length, or one or more values related to a particular setup situation.

In some manuals or texts, cutter diameter compensation and cutter length compensation are quite often called tool offset, cutter offset, tool length offset, etc.

In this chapter we will limit ourselves to the following uses of tool offsets:

6.1—Tool offsets used for length compensation
6.2—Flexible positioning of holding fixtures or parts
6.3—Multiple part machining (same or different)
6.4—Limited use in diameter compensation

These features are common to a large number of types of machines. The specific offsetting related to the ability to program the part rather than the tool will be called:

- Cutter diameter compensation (including the TNR compensation) and,
- Cutter length compensation.

In these specific cases the term "compensation" will be used consistently in lieu of "offset," and detailed discussion of these important features will take place in chapters 7 and 8.

6.1 TOOL OFFSET CODES USED FOR TOOL LENGTH COMPENSATION

Tool offset codes for tool length compensation are a very useful CNC programming feature that allow the operator to perform milling, drilling, tapping, or boring without presetting the tools to a specific length. As the name implies, this compensation controls the tool or "Z-" axis of a CNC machining center. The length of the compensation is controlled by the value stored by the operator in a specific offset register programmed for a particular tool.

It is a good technique to program a separate register for each tool length used. If the same tool is to be used in a different portion of the program requiring a different length, another register should be used.

The basic principle is similar, in many cases identical, for most controls now being built.

Tool offsets used in tool length compensation essentially represent *addition* or *subtraction* by the CNC control. In the process of drilling, we program a specific rapid "Z-" motion to approach the part surface. In addition, we program a tool register number, such as H12, and a tool offset code G45 for lengthening (adding).

N0040 G00 G45 Z-3.0 H12

The interpretation of this tape block is the following:

- N0040 —Sequence Number
- G00   —Rapid Motion
- G45   —Addition (tool offset)
- Z-3.0 —Programmed Length of Tool Motion
- H12   —Assigned Tool Register

This example is illustrated in the lower half of Fig. 6-1.

We shall assume that the operator entered 2.9 into H12. The drill point will move in rapid motion by the amount of programmed "Z" *plus* the distance contained in tool register H12. It should be remembered that the tool will move by the *amount* programmed in "Z," in the *direction* given by the sign in "Z." When calculating, we add to the unsigned amount from "Z" the signed amount from the offset register. The resulting motion will take place in the direction of the sign resulting by

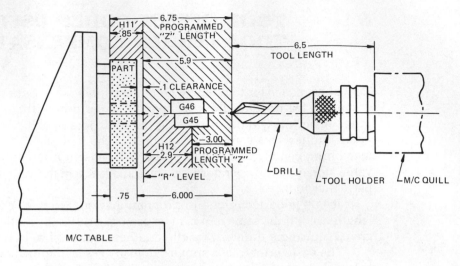

FIG. 6-1. Tool offsets used in tool length compensation.

combining the sign from "Z" with the sign of the added amounts. For example:

Z-3.0	H12 contains 2.9;	$3 + 2.9 = 5.9$; motion -5.9
Z-3.0	H12 contains -2.9;	$3 - 2.9 = 0.1$; motion -0.1
Z-3.0	H12 contains 3.9;	$3 + 3.9 = 6.9$; motion -6.9
Z-3.0	H12 contains -3.9;	$3 - 3.9 = 0.9$; motion $+0.9$

The tool offset code G46 will shorten (subtract) from the programmed motion the amount entered in the appropriate offset register.

N0050 G00 G46 Z-6.75 H11

Illustrated in the upper part of Fig. 6-1, this example can be interpreted as follows:

The drill point will move in rapid motion by the programmed amount "Z" or 6.75 inches, *less* the distance 0.85 inch assumed entered in tool register H11.

In actual case, both programmed tape blocks in the examples above will result in the same tool point motion to "R" level, or 5.9 inches.

Even though the system accepts both tool lengthening or shortening codes, most programmers set a standard of using either one or the other, but not both. This will avoid potential confusions that could result in expensive tool crashes.

Using G45 in our example means that the operator can use any tool length less than or equal to 6.5 inches + 2.9 inches = 9.4 inches. At 9.4 inches of tool length, the value of H12 would be zero without having had to change the programmed Z-3.0 dimension.

Using G46 on the other hand, means that the operator can use any tool length less than or equal to 6.5 inches + 5.9 inches = 12.4 inches. At 7.5 inches of tool length, the value of H11 would be 1.85 inches, and at 10.5 inches of tool length the value of H11 would have to be changed to 4.85 inches. In other words, whatever we add to the tool length we have to add to the content of the tool register.

By now the reader may have concluded that it is easier to work with the G45 code. Let us assume that the shortest possible tool length (i.e., cutting tool and holder combination) is 6.5 inches. The clearance between tool and part at the end of the rapid approach is, for example 0.1 inch. If we program a rapid motion of 1.0 inch, in conjunction with G45 and a particular tool offset register, the amount to be entered by the operator in that register will be the following:

The distance from the tool point to the part surface,
Less the programmed 1.0-inch motion,
Less the 0.1 inch-clearance.

6.2   TOOL OFFSETS USED FOR POSITIONING OF FIXTURE OR PART

Figure 6-2 illustrates a setup that requires the use of three different sizes of drills. Since we know how to write a part program using tool length compensation, we can ignore the actual length of our tools during the programming process.

The following steps will outline the job:

1. The three drills specified are mounted in three different tool holders. The lengths are unimportant from a programming point of view, but rigidity, deflection, and proper grinding remain important and should not be overlooked.

2. The plate shown in Fig. 6-2 is mounted on the machine table. The exact location is unimportant since we will use tool offsets in the X-Y plane. It is important to maintain the part edge marked "A" parallel to the table surface. The risers should be located so as to avoid tool interference.

3. The three tools are installed into the tool changer in the positions outlined by Fig. 6-2.

4. The distances between the tool tip and the part surface are found as follows:

- Position tool No. 1 in spindle.
- Zero "Z" axis in "home" position above part.
- Start spindle @ 100 rpm.
- Jog tool tip to touch the part.
- Read and note "Z" displacement. For our program, this reading is labeled Z1, and we will assume the respective amount to be 7.9 inches. The same exercise for tools Nos. 2 and 3 will yield Z2 = 6.879 and Z3 = 5.695.

5. The program for the drilling of the three holes shown in Fig. 6-2 is now written below:

N0010 G20 G40 G80 G91	Initialization
N0020 G92 X0 Y0 Z0	"Zero" control
N0030 S1200 M03	Spindle on, 1200 rpm
N0040 G45 G00 X1.0 D01	
N0050 G45 Y1.0 D02	

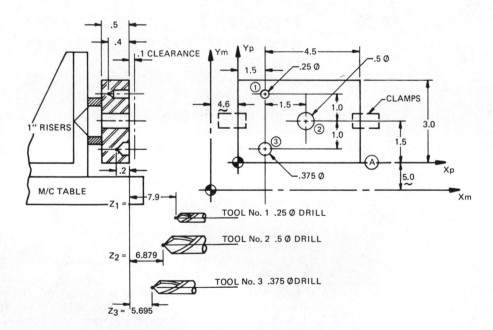

FIG. 6-2. Sample part for tool offsets.

Fig. 6-2 shows two sets of coordinates. The origin of Xm and Ym is the machine origin, while the intersection of Xp and Yp is the part origin. From the machine origin, the coordinates of point "1" can be calculated from the drawing as $X = 6.1$ inches and $Y = 7.5$ inches. Had these dimensions been used in the program, the operator would have been required to set the part exactly to the dimensions X4.6000 and Y5.0000. As it stands, the tool offset registers D01 and D02 have been used. The operator has the choice of positioning the part any place he wishes, as long as the edge "A" is parallel to the table. Once the part is positioned, the operator measures the X- and Y-dimensions of the actual location. Since a displacement of 1.0 inch has already been programmed in sequences N0040 and N0050, the operator will subtract 1 inch from the X-amount and will enter the result in D01. He will subtract 1 inch from the Y-amount and will enter this result in D02. Due to G45, the amounts in the offset registers combined with the programmed motions will bring the tool to the required location.

If we assume that the part has been mounted at the operator's convenience and that the dimensions measured from the machine origin to the part origin are, respectively, 4.6 and 5.0 inches, as shown on the drawing, we have the following choices:

Along X,

$$4.6 + 1.5 = 6.1$$

Since we have programmed X1.0, we enter 5.1 in D01
Along Y,

$$5.0 + 1.5 + 1.0 = 7.5$$

We enter 6.5 in D02.
Alternatively, we could have programmed:

N0040 G45 G00 X1.5 D01
N0050 G45 Y2.5 D02

The operator measures 4.6 and 5.0 inches as before and enters them directly as measured in D01 and D02. The results are the same. We can now continue the program in the Z-axis.

N0060 G00 G45 Z-1.0 H03   Tool length compensation

The value of the H03 offset register must at this point be calculated as:
$Z1 - 1.0$ inch $- 0.1$ inch or 7.9 inches $- 1.0$ inch $- 0.1$ inch $= 6.8000$

where 1.0 inch is the Z-value programmed in sequence N0060 and 0.1 inch is the clearance.

N0070 G01 Z-0.5 F3.0 M08	Drill hole 1, 0.250-inch diameter, 0.4-inch deep
N0080 G00 Z0.5	Rapid return
N0090 G28 Z0	Return tool home
N0100 X1.5 Y-1.0 M05	Rapid to 2nd hole
N0110 M06 T02	Tool change
N0120 S900 M03	
N0130 G45 Z-1.0 H04	Tool length compensation

The value of the H04 offset register must be calculated exactly as in the case of H03 above.

N0140 G01 Z-0.63 F400	Drill 0.50 inch thru
N0150 G00 Z0.63	Rapid return
N0160 G28 Z0 M05	Return tool home
N0170 X-1.5 Y-1.0	Rapid to 3rd hole
N0180 M06 T03	Tool change
N0190 S1050 M03	
N0200 G45 Z-1.0 H05	Tool length compensation

The value of the H05 offset register must also be calculated as above.

N0210 G01 Z-0.3 F350
N0220 G00 Z0.3
N0230 G28 Z0 M05
N0240 G28 X 0 Y 0 M09
N0250 M06
N0260 M30

6.3   TOOL OFFSETS USED IN MULTIPLE PART MACHINING

This paragraph represents an extension of the preceding one.

Let us assume that we're machining the part in Fig. 6-2, but that in lieu of having positioned the part using tool offsets, we have positioned a holding fixture for that part. We are now machining a small production lot.

In a jobbing shop, where production flexibility is vitally important, we have the ability to locate a second holding fixture in a different place on the table. The first fixture was located using D01 and D02. We can locate the second one using different offset registers, such as D11 and D12. We can now interrupt the first run, machine the second order, and then resume production. Where this type of requirement is a frequent occurrence, it is a good practice to program the X- and Y-motions first, with Z retracted all the way. This will avoid having to worry about crashing the tools for the second part while crossing the zone of the first one.

6.4 TOOL OFFSETS USED IN DIAMETER COMPENSATION

It is not recommended that offsets such as G45 or G46 be used in cutter diameter compensation. These codes will function in limited applications, for very simple geometries. Because errors may be introduced due to the design of the software, it is preferable to use G41 and G42 which have been specifically designed for this purpose. On specific controls, G41 and G42 may be used in conjunction with an additional code, G39. Cutter diameter compensation codes and their applications will be discussed in subsequent chapters.

Cutter Diameter Compensation and TNR Compensation: Programming the Work Surface

7.1   CUTTER DIAMETER COMPENSATION

We have seen the complexities of cutter center programming in chapter 5. While not difficult as such, programming the cutter center (sometimes called "programming the tool") can become involved, and the probability of errors increases with the number of calculations.

We shall now look at an entirely different approach. This consists of "programming the part." The programmer will write a part program tailored to the contour to be machined. This is equivalent to using an imaginary tool that has zero radius.

As part of the program, the programmer incorporates a "compensation" tape block. While the compensation "offsets" the tool, to minimize potential confusion in terminology, the process discussed in this chapter is "cutter diameter compensation" and not "tool offset." However, the numbered memory compartment where the value of the cutter radius will be stored is called "offset register." Offset registers are used in conjunction with different procedures such as tool offset, cutter diameter compensation, tool nose radius (TNR) compensation or tool length compensation.

The compensation tape block has the following functions:

1. To assign a specific offset register to a specific tool. CNC systems are normally equipped with programmable tool registers, available usually in blocks of 16; that is, a control may have 16, 32, 64, or more registers. Prior to running the program, the operator has to input (other terms used are "punch" or "dial") the desired compensation into the assigned offset register. This compensation is usually the radius of the tool.

2. To simplify the programming process in terms of roughing or finishing operations. Upon instruction, the operator can input in the offset register a radius that is larger than the actual radius of the tool used. The amount of the difference will represent the stock left on the part after the current pass.

3. To allow the use of a cutter of a different diameter than specified should a specific size become unavailable due to breakage, etc.

4. To define the direction of the compensation, to the right or to the left of the part.

5. To define the direction of the machining following the tool compensation block, i.e., conventional or climb milling.

The basic functions described above are by and large the same for most modern CNC systems. The tape block format may change, however, from one system to another. If a programmer acquires a thorough knowledge and understanding of the compensation process in one system, there should be no difficulty in switching this knowledge to a somewhat different format. Having looked at tool compensation in general terms, we will study several programs to bring the discussion to a more specific level.

7.1.1   Cutter Diameter Compensation Left—G41 (Example in Inches)

The compensation feature will position the tool, prior to machining, to the left of the part, by a distance equivalent to the tool radius input in the assigned tool offset register. The part program will be written in a very simplified cutter centerline mode, as if the tool radius were zero. The dimensions to be used for cutter motions will therefore become the actual dimensions of the part drawing, which precludes the need for calculations.

The part shown in Fig. 7-1 will be machined with a 0.50 inch-diameter-end mill, in conjunction with offset register No. 03 (shown in the program as D03).

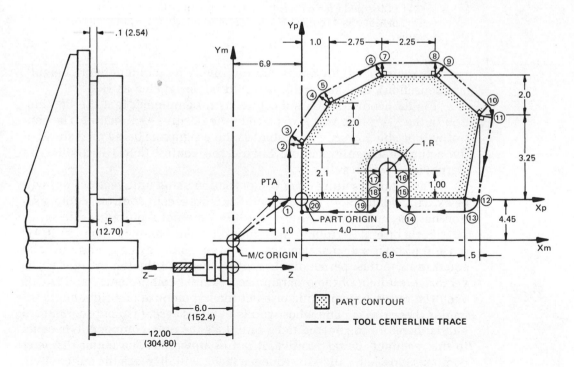

FIG. 7-1. Cutter diameter compensation.

The Direct Approach

Using this method, the tool is programmed to move directly to point 1. The implication here is that the operator must set the holding fixture on the machine table exactly to the dimensions given in the setup drawing.

```
N0010 G20 G40 G80
N0020 G92 X0 Y0 Z0
N0030 S900 M03
N0040 G91 G17 G00 G41 X6.9 Y4.45 J2.1 D03
```

Sequence 0040 will "rapid" the cutter from the machine origin to point 1.

- G91 represents incremental programming
- G17 indicates that compensation takes place in the X-Y plane
- G00 is rapid traverse
- G41 is the code for cutter diameter compensation left
- D03 is tool offset register No. 03, earmarked for X-Y plane programming.

As shown later, for Z-axis motion (involving tool length or "height" compensation), the same register would be designated as H03.

J is the word address indicating, as programmed, that the first motion in the X-Y plane will take place in the positive Y-direction. The control will be able to verify this intent as the program is being read ahead of the actual machine execution. Should the control detect a conflict, an alarm signal will result.

The detailed setting up of compensation using unit vectors I or J will be explained later on in this chapter. For more recent controls, compensation has been greatly simplified, and I or J are not required at all.

For a brief description, the cutter diameter compensation left (G41) or right (G42) is a displacement of the cutter center, by an amount usually equal to its radius, perpendicular, respectively, to the left or right to the vectorial resultant of the programmed vectorial components I and J. This vectorial resultant is nothing else but the hypotenuse of a right-angle triangle whose rise is J and whose run is I. In this case, I is not programmed and is therefore considered to be zero. J, as the only component, is equal to the resultant. Being positive, it points upwards. The cutter diameter compensation left, called in sequence 0040, will displace the cutter 90° to the left, as shown in Fig. 7-1.

N0050 G18 G00 Z-5.3 — Rapid down to 0.1 inch clearance above part surface

N0060 G18 G01 Z-0.65 F2.0 — Feed down to cutting depth

The next block will have to specify a positive Y-motion, as required by sequence N0040.

The Indirect Approach

This method is used by experienced programmers and is far superior to the preceding one. As described previously in chapter 6, the operator has the choice to position the holding fixture at any convenient location on the machine table, as long as it is set parallel to the machine axes. We arbitrarily select a point A, such as 1 inch away from the part, as shown in Fig. 7-1. This point will be reached using G45 tool offsets.

The preceding program would now read:

N0010 G20 G40 G80
N0020 G92 X0 Y0 Z0
N0030 S900 M03
N0032 G45 G00 X0 D01
N0034 G45 Y0 D02
N0040 G91 G17 G01 G41 X1.0 Y0 J2.1 D03
N0050 G18 G00 Z-5.3
N0060 G18 G01 Z-0.65 F2.0

Sequences 0010 to 0030 and 0050 to 0060 are unchanged, as is the rest of the program. The difference consists of sequences 0032 and 0034, where the operator will have to have input, prior to the start of machining, the location of point A. If we assume that the part shown in Fig. 7-1 has been located on the machine and its "part origin" measured at X6.9 and Y4.45, the operator will input 6.9 inches − 1.0 inch = 5.9 inches in register D01 and 4.45 in register D02. The use of two blocks for this procedure gives the operator additional flexibility.

From here on, the program will be identical for both direct and indirect approaches.

N0070 G17 G01 Y2.1 F3.0 M08 — Motion from point 1 to point 2

Most compensation systems operate in two ways, depending on their executive software. The older systems require an additional code, G39, whose purpose is to rotate the tool centerpoint, as shown in the transition from points 2 to 3, 4 to 5, 6 to 7, etc., on Fig. 7-1. This occurs in conjunction with a change in direction of the linear interpolation, and it is required to bring the tool radius perpendicular to the new surface to be machined. As illustrated in chapter 5, regardless of the programming method, the cutter radius will be perpendicular at all times to the surface being machined (which in its turn is permanently "tangent" to the cutter "circle"). In the newer systems, as indicated by their manuals, the CNC software will take care of this problem by "automatically" bringing the cutter center to the intermediate position where the cutter "circle" will be "tangent" to both directions, the current one being machined, and the next one in line. For the present program, G39 will be used, as illustrated in Fig. 7-2.

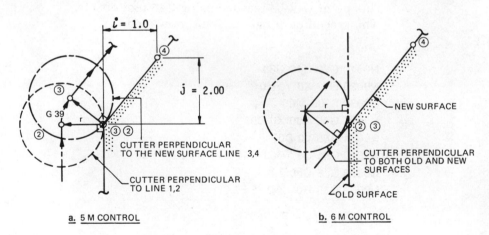

FIG. 7-2. Tool radius vector setting.

At the end of block 0070, the tool radius is perpendicular to part surface 1-2, between points 1 and 2. Prior to programming surface 3-4, we require a tape block to rotate the tool at point 2, so that its radius becomes perpendicular to the next surface, 3-4. It can be seen in Fig. 7-2a that should this rotation not occur, the new surface will be cut to the right of the "true" line 3-4 and the part will end up smaller and scrapped.

N0080 G39 I1.0 J2.0

This tape block performs the reorientation of the tool radius. G39 is a pre-programmed subroutine, stored in control memory. The values for I and J have been selected as follows: $I = 1.0$, which corresponds to the X-displacement from point 3 to point 4 as seen in Fig. 7-1. $J = 2.0$ is, in the same sketch, the corresponding Y-displacement.

I and J can be interpreted as the two component vectors whose resultant is line 3-4, or as the two sides of a right-angle triangle whose hypotenuse is line 3-4. The two definitions are geometrically identical. As G41 called in sequence 0040 for compensation "left," our tool radius will now be oriented 90° to the new line 3-4, defined by I and J, as shown in N0080.

What the CNC minicomputer calculates is the slope of line 3-4, given by its rise J, divided by the run I. As the result is a ratio, we could have programmed in N0080 any pair of values for I and J whose ratio is the same. We could have had, for instance:

```
N0080 G39 I10 J20 or
N0080 G39 I100 J200 or
N0080 G39 I50 J100, etc.
```

However, it is recommended as much as possible to use pairs of values, as found on Fig. 7-1, to simplify possible debugging. The G39 vector setting code must be programmed prior to each change of direction.

As seen in Fig. 7-2b, illustrated for a 6M control, the motion programmed in sequence 0070 will bring the cutter center to a position where the tool radius, while still perpendicular to the "old" surface, becomes perpendicular to the "new" surface as well. The software of the newer controls will automatically perform calculations and generate motions without the use of G39, or the extensive calculations from chapter 5.

Continuing the program,

```
N0090 G01 X1.0 Y2.0      From point 3 to point 4
N0100 G39 I2.75 J1.15    Rotation around point 4
N0110 G01 X2.75 Y1.15    From point 5 to point 6
N0120 G39 I2.25          Rotation around point 6
```

As line 7-8 is parallel to the X-axis, sequence 0120 requires no J and the next one, 0130, requires no Y.

N0130 G01 X2.25	From point 7 to point 8
N0140 G39 I1.4 J-2.0	Rotation around point 8
N0150 G01 X1.4 Y-2.0	From point 9 to point 10
N0160 G39 I-0.5 J-3.25	Rotation around point 10
N0170 G01 X-0.5 Y-3.25	From point 11 to point 12
N0180 G39 I-0.9	Rotation around point 12
N0190 G01 X-0.9	From point 13 to point 14
N0200 G39 J1.0	Rotation around point 14
N0210 G01 Y1.0	From point 15 to point 16
N0220 G03 X-2.0 I-1.0	Circular interpolation

G02 or G03 will gradually rotate the tool radius as it machines the part surface, therefore, no G39 is required at either point 16 or 17.

N0230 G01 Y-1.0	From point 17 to point 18
N0240 G39 I-3.0	Rotation around point 18
N0250 G01 X-3.0	From point 19 to point 20
N0260 G18 G00 Z0.65	Raise tool to 0.1-inch clearance
N0270 G17 G40 X-1.0 Y0	Cancel cutter diameter compensation
N0280 G28 Z0	Return Z "home"
N0290 G28 X0 Y0	Return X and Y
N0300 M30	End

Briefly, this process described above saves calculation time and eliminates calculation errors.

7.1.2 Cutter Diameter Compensation Left (Metric Example)

The sample part shown in Fig. 7-3 is dimensioned in metric. As an observation, dimensions on metric blueprints may be shown without trailing zeros, as their dimensional tolerances are not normally related to the number of zeros after the decimal point.

The only significant change is the G21, replacing G20, which indicates to the control that the following dimensional values are given and should be reflected in metric dimensions. The inch/metric is usually programmed G20/G21 on some controls, and G70/G71 on others. When a

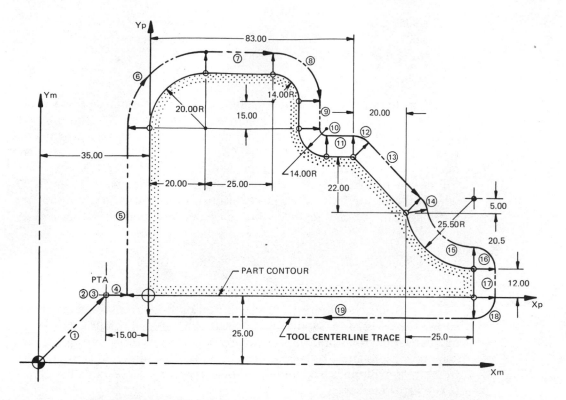

FIG. 7-3. Cutter diameter compensation—left (metric part).

CNC system works only in metric or only in inches, the above codes are not required.

    N0010 G21 G40 G80 G91

As G41 or G42 set up a cutter diameter compensation condition, G40 is the code that cancels it. An uncanceled compensation can generate a lot of problems, therefore, G40 is always used in the first block as a safety feature. G80 performs the same function for canned cycles, to be discussed in detail later, in chapter 9.

    N0020 G92 X0 Y0 Z0
    N0030 S800 M03
    N0040 G45 G00 X0 D01   Rapid to point A using G45
    N0050 G45 Y0 D02

N0060 G18 Z-302.0	Rapid to clearance
N0070 G18 G01 Z-16.0 F10.0	Feed to cutting level
N0080 G91 G17 G01 G41 X15.0 Y0 J68.5 D03	

For a control without the decimal point programming feature, the tape word X15.0 would be programmed X1500 for an X5.2 tape format (which means that the largest programmable dimensional word on the respective system would be X99999.99, or 5 digits before and 2 after the decimal point—a metric value representing just over 328 ft.). For a control of this nature, it may be helpful to mark the dimension 15 as 15.00, which allows for its direct transposition into the program, without the decimal point, of course, but with the correct number of trailing zeros.

N0090 G17 G01 Y68.5 F2.5 M08	Straight-line milling of zone 5

As a point of interest, one of the few applications of the "plane" codes, G17, G18, and G19 occurs in cutter diameter compensation. As the compensation takes place in the X-Y plane (G17), unless the programmed Z-motion is preceded by a G18, the control will assume that the departure from the X-Y plane is accidental and the software will generate the appropriate error message.

N0100 G02 X20.0 Y20.0 I20.0	Circular interpolation, zone 6
N0110 G01 X25.0	Linear interpolation, zone 7
N0120 G02 X14.0 Y-14.0 J-14.0	Zone 8
N0130 G01 Y-15.0	Zone 9
N0140 G03 X14.0 Y-14.0 I14.0	Zone 10
N0150 G01 X10.0	Zone 11
N0160 G39 I20.0 J-22.0	Rotation of tool, zone 12
N0170 G01 X20.0 Y-22.0	Zone 13

For the cutter repositioning at zone 14 (Fig. 7-4), it may be helpful to sketch an enlarged view of the part geometry in zone 15.

The cutter radius must rotate as shown at point 14 from an upper position perpendicular to zone 13, to a lower position perpendicular to zone 15. At this lower position, the cutter radius must be collinear with the radius R of zone 15. This rotated position of the tool radius must be at 90° to the left of the hypotenuse (R) of the right-angle triangle whose sides are the I and J programmed with G39.

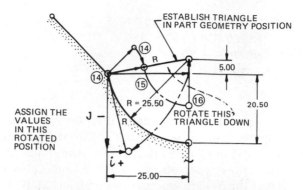

FIG. 7-4. Unit vectors at rotated position.

- Draw a line from point 14 to the center of arc (R)
- From point 14 draw a horizontal line (dimension 25) and a vertical one (dimension 5)
- The three lengths, R, 25, and 5 form a right-angle triangle
- At this stage, R is collinear with the rotated position of the tool radius
- Rotate the whole triangle as shown, 90° clockwise. R is now perpendicular to the short arrow representing the cutter radius. Looking in direction of the arrow of R, the cutter radius arrow points to the left, according to G41. Therefore, R is in correct position, and the triangle side parallel with the X-axis is I (shown on Fig. 7-4 as i, positive), while the other side is J (shown as j, negative).
- The rotation of the triangle has not changed the values of its sides, it has only helped to assign correct signs.

Accordingly,

N0180 G39 I5.0 J-25.0   Rotation, zone 14
N0190 G03 X25.0 Y-20.5 I25.0 J5.0   Zone 15

It should be observed that as G39 precedes a circular interpolation, its *I* and *J* values are different from the ones in the subsequent motion statements, and not the same, as we had observed repeatedly in the previous program.

N0200 G39 J-12.0   Rotation, zone 16
N0210 G01 Y-12.0   Zone 17
N0220 G39 I-20.0   Rotation, zone 18
N0230 G01 X-128.0   Zone 19
N0240 G18 G00 Z16.0   Tool up

N0250 G17 G40 X-15.0 M05   Cancel compensation
N0260 G28 Z0 M09
N0270 G28 X0 Y0
N0280 M30

As a closing comment, it is always advisable to set up and cancel compensation in conjunction with a straight-line motion.

Most modern controls allow simultaneous return "home" on three axes. Unless your specific control completes the Z-motion prior to the start of the X-Y, in conjunction with a G28, it is recommended to program as in the above example, to assure that the tool is out of the way prior to any side motion.

7.1.3   Cutter Diameter Compensation Right—G42

For simplicity, the tool length and part setup will be assumed the same as for the two preceding programs. See Fig. 7-5.

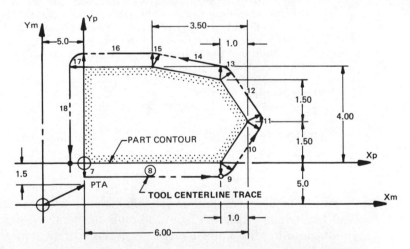

FIG. 7-5. Cutter diameter compensation right—G42.

N0010 G20 G40 G80 G91
N0020 G92 X0 Y0 Z0
N0030 S900 M03
N0040 G00 G45 X0 D01
N0050 G45 Y0 D02   To point A

```
N0060 G18 Z-5.3
N0070 G18 G01 Z-0.65 F3.0
N0080 G17 G01 G42 X0 Y1.5 I5.0 D03
N0090 G01 X5.0 F3.5 M08                    Zone  8
N0100 G39 I1.0 J1.5                               9
N0110 G01 X1.0 Y1.5                               10
N0120 G39 I-1.0 J1.5                              11
N0130 G01 X-1.0 Y1.5                              12
N0140 G39 I-2.5 J1.0                              13
N0150 G01 X-2.5 Y1.0                              14
N0160 G39 I-2.5                                   15
N0170 G01 X-2.5                                   16
N0180 G39 J-4.0                                   17
N0190 G01 Y-4.0                                   18
N0200 G18 G00 Z0.65 M09
N0210 G17 G40 Y-1.5 M05
N0220 G28 Z0
N0230 G28 X0 Y0
N0240 M30
```

7.2 TOOL NOSE RADIUS COMPENSATION

In principle, the turning center tool nose radius is no different from the milling cutter radius. In fact, though, it is much smaller, and it does not turn. Accordingly, we must look at a "standard tool nose vector." A vector is a quantity defined by its direction, or orientation, and its magnitude, or size.

The "direction" is usually shown as an arrow from the center of the TNR to the intersecting point of the two cutting lines of the tool.

The "magnitude" is equal to the tool nose radius when the compensation mode is activated. It is lengthened to the intersection of the two cutting lines when compensation is canceled.

In milling, the magnitude ranged from zero, i.e., cutter center, to the cutter radius, i.e., the cutter edge.

The vector of a standard tool nose is viewed from behind the tool during turning, in order to correctly assign the two directions, left or right. For example, the Fanuc 6T control provides eight different types of tool geometries, as shown in Fig. 7-6. Depending on which cutter geome-

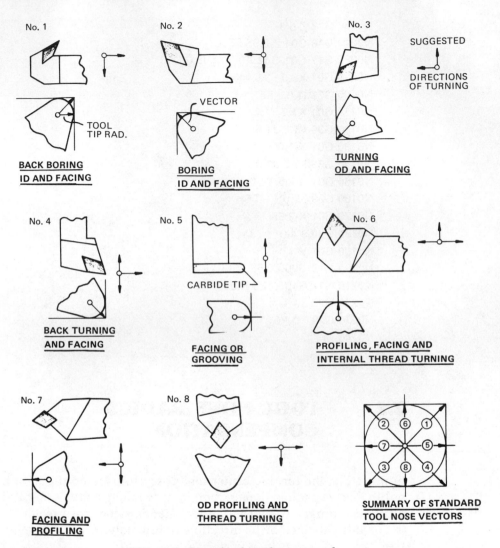

FIG. 7-6. Standard tool nose tools.

try suits a particular application, the programmer must also select this standard tool nose number together with its corresponding code.

The TNR and the number of the standard tool nose must be input into the respective tool offset register prior to running the program, as shown in Table 7-1.

Turning centers usually have 16 or 32 programmable tool offset registers. These can be shown in the program in the tool word as follows:

e.g., N25 T0212

TABLE 7-1 TOOL OFFSET SHEET				
Offset register number	Compensations		Tool nose radius R	Standard tool number
	X	Z		
12	−0.006	0.009	0.031	4

The first two digits following the "T" refer to the tool turret position. The second pair of two digits represent the tool offset register number.

TNR compensation left, G41, when programmed, will lead to the position shown in Fig. 7-7:

- The tool tip is to the left of the surface already machined.
- The direction of the tool parallel to the turned surface is to the left of the vector perpendicular to the machine surface.

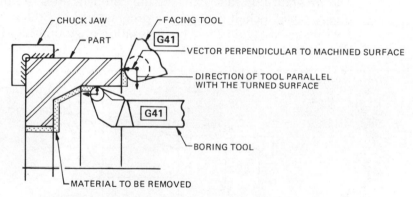

FIG. 7-7. Tool nose radius compensation left—G41.

This compensation condition is maintained by the control unit until a compensation cancel code, G40, is programmed. The G41 code is most often used for internal turning and facing.

TNR compensation right, G42, when programmed, will position the tool tip as shown in Fig. 7-8:

- The tool tip is to the right of the surface already machined, as we look at the turned surface, along the vector perpendicular to it.
- The direction of the tool parallel to the turned surface is to the right of the vector perpendicular to the machined surface, as we look along it, from its tail to its tip.

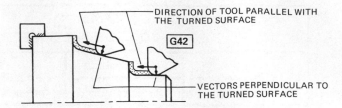

FIG. 7-8. Tool nose radius compensation right—G42.

TNR compensation right is used for contour turning of outside diameters.

In both cases, left and right, the control will maintain the vector perpendicular to the turned surface, regardless of the type of contour.

7.2.1 Setting Up Tool Nose Radius Compensation

The program line that activates the compensation is called the start-up line or tape block. It can only be programmed in G00 or G01 modes. Fig. 7-9 illustrates a typical tool compensation programming situation for both G41 and G42.

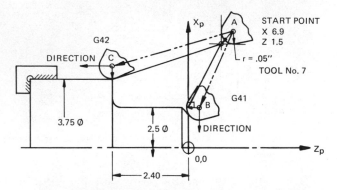

FIG. 7-9. Start up in tool compensation mode.

Assuming the tool to be at the start point "A" for both cases, we can ignore the tool dimensions during programming, since we are programming the part, not the tool. The tool dimensions will have to be input in tool offset register No. 7 prior to turning the part.

N0010 G00 X2.5 Z0.5 T0707 Rapid to clear part
N0020 G01 G41 Z0 F0.05   Compensation at "B"

or

     N0010 G00 X4.0 Z-2.4 T0707   Rapid to clear part
     N0020 G01 G42 X3.75 F0.05   Compensation at "C"

In both cases the tool compensation was set from a short, safe distance away from the part, to save machining time. As well, both applications show machining of surfaces parallel to the primary axes of the coordinate system.

7.2.2   *Canceling the Compensation—G40*

Canceling the compensation—G40 is the reverse process of setting the tool compensation. We shall assume that the turning was finished at points "*B*," respectively "*C*," and it is desired to cancel the tool compensation condition. It is good practice to move the tool to a safe location away from the part before the G40 code is programmed.

     N0030 G01 Z0.5 F35.0   Tool away from part
     N0040 G40 G00 X6.9 Z1.5   Cancel compensation

These two blocks could also be written as follows:

     N0030 G01 X4.0 F35.0   Tool away from part
     N0040 G40 G00 X6.9 Z1.5   Cancel compensation

This safe method of programming should be used until one becomes reasonably certain that the tool will move away from the part in the compensation cancel block. The two above versions could be programmed, omitting sequence N0030.

When compensation is canceled directly off the part surface, the control provides a special programming feature for tapered surfaces, using unit vectors as shown in Fig. 7-10.

Assume the last tool motion programmed to be from point *A* to point *B*.

     N0050 Z-1.75

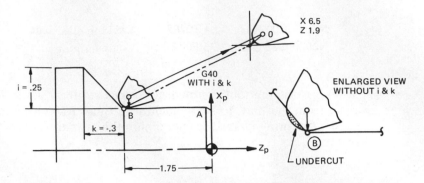

FIG. 7-10. Tool compensation using unit vectors i and k.

At this point, the tool tip center is under compensation control, and its radius is tangent to both cylindrical and tapered surfaces at the end of the motion.

If at this point the tool compensation is canceled, prior to returning to point 0, we must program in the same block an "I" and a "K" in order to prevent the tool center point from jumping in line (vertically) with point B. This would result in an undercutting position, as illustrated in Fig. 7-10. The correct programming procedure is shown below in sequence N0060.

N0060 G40 G00 X6.5 Z1.9 I0.25 K-0.3 T0700

It should also be borne in mind that while the compensation feature is on, each and every line of program must contain a motion. No tape blocks should be inserted that contain only M, S, G codes, or time delays, because the result will be an undercutting condition.

7.2.3   Tool Nose Radius Compensation Left—G41

As mentioned earlier, TNR compensation left—G41 is normally used for internal turning. In Fig. 7-11, the tool will rapid to a safe clearance of 0.5 inch. The compensation will be set up in conjunction with a G01 motion.

To simplify the program and keep it to a manageable size, it shall be assumed that the part was already rough turned.

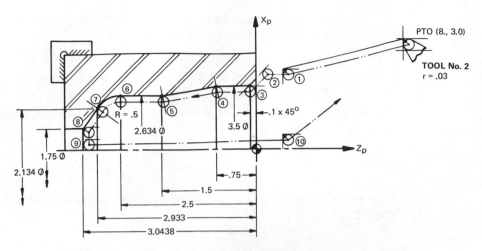

FIG. 7-11. Tool nose radius compensation left—G41.

N0010 M41	2nd gear range
N0020 G50 X0 Y0 Z0	"Zero" machine in home position
N0030 G99	Feed in ipr
N0040 G00 X-5.3 Z-13.0	Rapid to point 0
N0050 G50 X8.0 Z3.0	Transfer coordinates to part
N0060 T0202 S1200 M03	Index tool
N0070 G00 X3.6 Z0.5	Rapid to point 1
N0080 G01 G41 X3.9 Z0.1 F0.06	Set compensation to point 2
N0090 X3.5 Z-0.1 F0.005 M08	To point 3
N0100 Z-0.75	To point 4
N0110 X2.634 Z-1.5	To point 5
N0120 Z-2.5	To point 6
N0130 G03 X2.134 Z-2.933 R0.5	Circular interpolation to point 7
N0140 G01 X1.75 Z-3.0438	To point 8
N0150 X1.5	To point 9
N0160 G00 G40 X1.4 Z0.5	To point 10

The X1.5 in sequence N0150 moves the tool point a safe distance away from the part prior to canceling the tool compensation. The alternate method is to program a negative K value, to prevent the tool from moving in the negative X-direction at point 8, which would result in an undercut.

In the second method, sequence N0150 becomes redundant, and the balance of the program is shown below:

N0160 G00 G40 X1.4 Z0.5 K-5.0	Cancel compensation
N0170 X8.0 Z3.0 T0200	Return to point 0
N0180 G50 X-5.3 Z-13.0 M09	Coordinates back to machine
N0190 G28 X0 Z-13.0 M05	Turret slide to "home"
N0200 M30	End of program

Tool offset register No. 02 will have to be set prior to running the program.

7.2.4   *Tool Nose Radius Compensation Right—G42*

Tool Nose Radius Compensation Right—G42 will be illustrated in the program below, written for the part shown in Fig. 7-12.

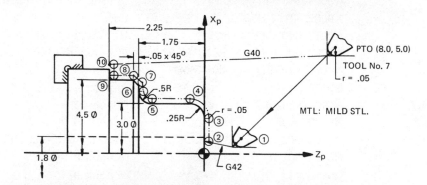

FIG. 7-12. Tool nose radius compensation right—G42.

N0010 M41	Second gear range
N0020 G50 X0 Z0	"Zero" machine in home position
N0030 G99	Feed in ipr
N0040 G00 X-4.0 Z-16.5	Rapid to point 0
N0050 G50 X8.0 Z5.0	Transfer coordinates to part
N0060 T0707 S1200 M03	Index tool

N0070 G00 X1.6 Z0.2	Rapid to point 1
N0080 G96 S680	Constant surface speed
N0090 G01 G42 X1.7 Z0 F0.05	Set compensation, motion to point 2
N0100 X2.5 F0.005 M08	Finish turn face to point 3
N0110 G03 X3.0 Z-0.25 R0.25	Contour to point 4
N0120 G01 Z-1.25	Turn 3.0-inch diameter to point 5
N0130 G02 X4.0 Z-1.75 R0.5	Contour to point 6
N0140 G01 X4.4	Face to point 7
N0150 X4.5 Z-1.8	Turn chamfer to point 8
N0160 Z-2.25	Turn 4.5-inch diameter to point 9
N0170 X5.1	Face, exit to point 10
N0180 G00 G40 X8.0 Z5.0 T0700	Cancel compensation
N0190 G50 G97 X-4.0 Z-16.5	Coordinates back to machine
N0200 G28 X0 Z0 M05	Turret slide to "home"
N0210 M30	End of program

Prior to running this program, we would have to insert the tool tip radius of 0.05 inch and the standard tool No. 3 into tool offset register No. 07, in manual data input (MDI) mode.

8

Tool
Length
Compensation

Tool length compensation is a code found on a number of machining centers equipped with more recent controls. It was designed with a specific purpose in mind, unlike the multipurpose tool offset, discussed in chapter 6.

Tool length compensation, as its name indicates, is used to compensate for tool length differences. It can perform this function in two directions:

1. G43, away from the part, or upwards in a vertical machine

2. G44, toward the part, or downwards if the machine is vertical.

As in other compensations previously discussed, cancelation is provided through a tool length compensation cancelation code, G49.

The major advantage of this feature is that it provides the programmer with the ability to write a complete program without knowing exactly the length of the tools to be used. The alternative is a set of very accurate calculations involving the distance from the quill face to the table; the dimensions of holding fixtures, spacers, risers, material thickness; the exact length of the holder-tool combination; and all the "Z" motions. These calculations would have to be performed by the programmer, rigorously implemented by the operator, and modified every time anything changes.

Tool length compensation introduces an elastic link in this chain of dimensions. The programmer can now write the entire program, including all the vertical motions, by using an imaginary tool length. This may be shorter or longer than all the tools the operator is likely to load in the automatic tool changer.

8.1 TOOL LENGTH COMPENSATION AWAY FROM THE PART—G43

In the case of a vertical machine, it is probably easier to visualize G43 as the tool length compensation "upwards," as this motion gets it away from the part surface.

The principle is the following: The programmer will assume that all the tools to be used are zero in length, i.e., he or she will program the quill face (adjusted for any gap that may exist on a particular machine between the quill face and the back of the tool holder). It should be noted that the reference to "tools" always indicates the holder-cutting tool combination.

Accordingly, should the distance from the quill face to the surface to be attained be, e.g., 15 inches, the appropriate program line would read:

```
N0010 M06 T01
N0020 G91 G00 G43 Z-15.0 H01
```

In tool offset register No. H01, prior to running the program, the operator will have inserted, in MDI mode, the true measured length of tool No. 1 (T01 from sequence 0010), e.g., 11 inches.

Consequently, the actual rapid motion of the spindle toward the part will be:

$$15.0 \text{ inches} - 11.0 \text{ inches} = 4.0 \text{ inches}$$

The major advantage of this approach is the time-saving for the programmer who no longer has to worry about tool lengths throughout the program. The increase in efficiency at the machine is also appreciable, since the setup will no longer have to be achieved to a set of given dimensions. The operator now makes the setup, measures it, and records the measurements. Should tool T01 have to be replaced later on by a reground version, different in length, the operator only has to update the information in tool offset register H01. The program remains unchanged, and the operator need not access it in memory should the lengths of the tools vary.

The potential pitfall is of course that the "cutterless" motions programmed are perforce fairly long, and should the operator overlook to input the tool length value, a serious crash would result in the Z-axis.

It can be seen that in the absence of a tool length compensating feature, the programmer would have to know exact tool lengths and program exact Z-motions. Any change in the length of any tool would result in a requirement for a program change. The compromise, in the absence of G43, is the use of the tool offsets, such as G45, discussed in chapter 6.

8.2   TOOL LENGTH COMPENSATION TOWARD THE PART—G44

The other version available with this feature is the tool length compensation toward the part, or "downwards" in a vertical machining center.

The principle here is the opposite to the previous "zero-length" assumption. The programmer will select an identical very large value for the length of all the tools to be used in the program. A suitable value is the nearest rounded-off value to the distance between the retracted quill face and the part surface, as shown in the following example:

Let the known (measured) distance between the table surface and the retracted quill face be 32.4000 inches. Assume that the measured distance between the table surface and the part surface to be machined is 3.2500 inches, including part thickness and fixturing. The calculated distance between the retracted quill and the part surface will be:

32.4000 inches − 3.2500 inches = 29.1500 inches

The amount of clearance required by the downward motion of the automatic tool changer is, e.g., 6.5000 inches.

29.1500 inches − 6.5000 = 22.6500 inches

The programmer can assume in the program that all the tools to be used are, e.g., 20.0000 inches long, and all the Z-motions will be programmed accordingly.

As the tools are quite "long," these Z-motions will be fairly "short."

The operator will set up the tools in accordance with the instruction sheet, will measure their exact length, will subtract the measured length from the value 20.0000 inches, supplied on the sheet, and will insert each difference in the respective tool offset register, as outlined by the Tool and Offset Sheet. The control will add these "differences" throughout the program, to the "short" motions specified by the programmer, thereby compensating for the difference in length between the actual values and the imaginary 20-inch tool.

Although this method entails an additional operation on the part of the operator, i.e., subtracting the length measured from the imaginary length supplied with the program, from a safety standpoint it is preferable to the other one. If the operator forgets to insert a value in the appropriate tool register, the motion will be short, but no crash will result.

Either length compensation method may be used, if available on the control. Once a method is adopted, there should be no deviations from it as they lead to confusion, and expensive crashes could result.

Tool length compensation may be used in conjunction with two types of motion:

- X-Y plane motions, as in milling
- Z-axis motions, as in drilling, tapping, or boring.

Where four-axis programming is involved, the control manual should be carefully checked as tool length compensation may be inoperable, or subject to a set of limitations.

In X-Y operations, it is advisable to set up the tool compensation in combination with a rapid (G00) motion to a safe clearance above the part. The balance of operations will take place in feed (linear or circular, as applicable).

In Z-axis operations, the compensation may be set up in conjunction with a rapid motion, to the "Initial level" of the subsequent canned cycle operation (see chapter 9). While the tool would still continue to the "R" level in rapid, this method would allow return to the "Initial level" without affecting the compensation condition.

8.3   TOOL LENGTH COMPENSATION CANCELATION—G49

Once the X-Y operations have been completed and the tool returned to the clearance location, a rapid return to start in Z, with a G49, will cancel the compensation. The same will apply to a succession of Z-axis operations, once the tool was returned in rapid to the "Initial level."

It is advised that the cancelation of tool length compensation take place in the same length of motion (but the opposite direction) as the initiating motion in G43 or G44.

As many other CNC features, tool length compensation is particularly appreciated by people who programmed and made parts on NC machines before it became available.

9

Canned Cycles

CANNED CYCLES may be defined as a set of preprogrammed instructions stored away in computer memory. The word "canned" has probably been borrowed from canned goods which one usually stores away for later use.

Because the instructions represented a set of routinelike repetitive patterns, the word "cycle" was found to best express what was taking place.

These canned cycles are filed away under a G-code address. To a large extent the G-codes are standardized.

As an example (see Fig. 9-1), a G84 code, usually representing right-hand tapping in a machining center, will consist of the following steps:

- Clockwise rotating of the tap at the correct rpm
- Rapid advancing of the tap to a set clearance from a predrilled hole (R-level)
- Feeding the rotating tap to a set depth at a rate of one thread pitch per revolution
- Reversing both feed and spindle rotation until the tap reaches the R-level.
- Returning spindle rotation to its original clockwise direction.

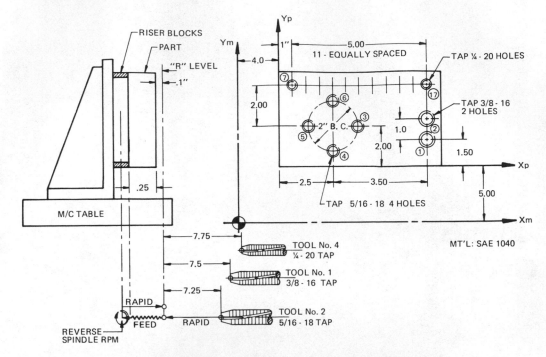

FIG. 9-1. Tapping canned cycle G84.

The spindle is now ready to rapidly return "home" for a tool change, relocate to another position or carry out any other instruction of the program.

Similar motion patterns are available for drilling, boring, turning, threading, etc.

MULTIPLE REPETITIVE CYCLES or AUTOMATIC REPEAT CYCLES: These represent the generating of complex profiling patterns in stock removal. They are very similar to canned cycles in that they are preprogrammed and prestored. Together the canned cycles and the multiple repetitive cycles are known as FIXED CANNED CYCLES.

VARIABLE CANNED CYCLES or SUBROUTINES: These emerged because of the increase in the number and complexity of prestored canned cycles and multiple repetitive cycles. This led to increases in the size of computer size and price, combined with slowdowns in the speed of processing.

The variable canned cycles allow the programmer to write a canned cycle to suit a particular requirement or application. A larger portion of the memory will now be available to handle a set of specific instructions, perform calculations, manipulate variables, carry out other assignments, at a far higher rate of speed.

9.1 FIXED CANNED CYCLE PROGRAMMING

9.1.1 Machining Centers, Vertical or Horizontal

As usual, the tool is always considered to be traveling in relation to a fixed part regardless of the fact that in one or more directions, the actual motion may be carried out by the machine table.

THE RIGHT HAND COORDINATE SYSTEM: With positive "Z" pointing toward the quill, this system will define the three axes of linear motion for a given machine, be it vertical or horizontal.

In the following discussions, we have selected a horizontal machine for our programs. We have also used DECIMAL POINT PROGRAMMING as its use is becoming more widespread. Thus a motion of 3.5 inches will be shown as X3.5 in lieu of the more conventional X35000. A motion of 2 inches would be shown then as Y2.0, or simply Y2., in lieu of Y20000.

The *general format* of the fixed canned cycle is the following:

N_ G_ G_ X_ Y_ Z_ R_ Q_ P_ F_ L_

where (see Fig. 9-2),

   N is the sequence number of the tape block
   G is the respective canned cycle (e.g., G81, G73, G76, etc.)
   The second G may be G98 or G99, to be discussed later
   X and Y are the coordinates of the location to be machined
   Z is the level attained by the tool in feed mode
   R is the level reached by the tool at the end of its rapid motion
   Q is the depth of cut in feed in peck drilling or the lateral shift in boring canned cycles
   P is the dwell in spot facing or boring operations
   F is the feed code
   L is the number of repeats, if applicable

(Codes not shown in Fig. 9-2 will be discussed in detail later in this chapter.)

The following observations apply to canned cycles for machining centers in general:

   • The tool motion starts in canned cycle from an INITIAL LEVEL
   • The tool will move in RAPID mode to the programmed X-Y location
   • The quill will move in RAPID mode to the R (for Rapid) LEVEL

- The tool will feed to the Z-LEVEL, measured from INITIAL in absolute mode (G90) and from R in incremental mode (G91)
- If the tool is already on location, there is no need to program X and Y
- G99 will return the tool point to R level; G98 will return the tool point to the INITIAL level
- A canned cycle is in effect until canceled (or replaced by another canned cycle). Accordingly, subsequent locations can be machined by simply programming X and Y in subsequent blocks. *If you do not intend* to perform the canned cycle at the next location, you must cancel it first
- A canned cycle can be programmed in either incremental or absolute mode
- Since X-Y motions are strictly for positioning and the programmed machining patterns take place in the Z-axis, no canned cycles are to be used in cutter diameter compensation mode
- In the examples shown below, tool offsets and cutter length compensation will be minimized or omitted in order to allow the reader to concentrate on the respective canned cycle process. Preset tooling will be used
- Most systems provide numerous canned cycles as standard equipment, but not all systems use the same codes to designate the same cycles. Always use the codes provided by your system manual.

In summary, therefore, there are three levels of tool position in any canned cycle:

- The INITIAL LEVEL which is the specific position of the tool at the moment the canned cycle becomes effective
- The "R" for RAPID level, which is the end of the rapid quill motion, usually 0.020 to 0.100 inches (0.5 to 2.5 mm) from the part surface, and
- The "Z" level which is the end of the feed (metal cutting) motion.

Usually, in one tape block, the canned cycle will control all or most of the following motions:

- Positioning the tool in the X-Y plane (Remember: the machine may position the part, but it is the tool motion that we program)
- Rapid motion to the "R" level
- Controlled feed motion to the "Z" level
- Commanded motions at "Z" level, such as dwell, spindle stop, spindle reverse, spindle orientation, spindle lateral shift, etc.
- Rapid or feed return to "R" level, as applicable
- Rapid return to INITIAL level.

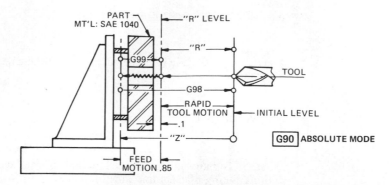

CANNED CYCLE IN ABSOLUTE MODE

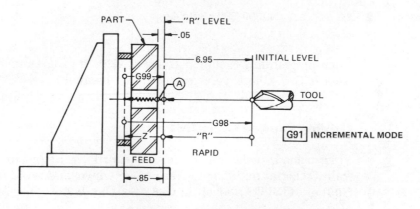

CANNED CYCLE IN INCREMENTAL MODE

FIG. 9-2. Canned cycle in:
   a) Absolute mode.
   b) Incremental mode.

Canned Cycle Cancelation G80

This code is normally used in three situations:

- When different programs are run consecutively on the same machine, it is used before the first motion statement, in conjunction with other codes as a safety feature, to cancel potential leftover canned cycles from a previous program. On some controls, a previously used canned cycle may remain active even if the control is reset or the power is turned off.

e.g., N0010 G20 G40 G80 G49 G91 G00

- When a canned cycle is canceled, as it should be, as soon as it is no longer needed in the program

e.g., N0040 G80

- When it is required to reposition the spindle without any machining taking place at that location (such as an intermediary position in a change of direction)

e.g., N0090 G80 X3.0 Y4.5

(Inch programming, shown above as G20, is programmed on many controls as G70. Check your programming manual.)

Drilling Canned Cycle G81

This canned cycle will perform the drilling of one or more holes as follows: Rapid to "R" level, feed to "Z" level and rapid return to "R" (in G99) or to INITIAL level (in G98). See Fig. 9-3.

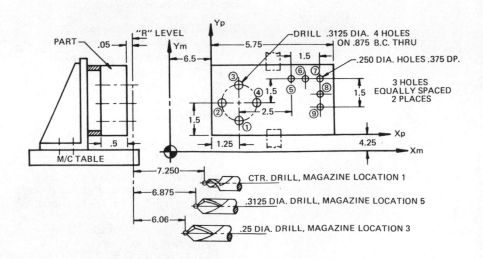

FIG. 9-3. Sample part for drilling canned cycle G81.

NOTE: *Comments or explanations placed on the program lines do not form part of the program. Some controls on the market allow the programmer to place data or instructions in brackets at the end of the program line. Such data will show on the CRT but will be overlooked by the machine. Other controls allow comments on a separate line, preceded by a specific G code.*

N0010 G20 G40 G80 G91 inch programming shown as either G20 or G70, cancel cutter diameter compensation, cancel canned cycles, incremental

N0020 B180 index table 180°

N0030 M08 coolant ON

N0040 M06 T01 tool change, select tool No. 1

N0050 S1500 spindle speed 1500 rpm

N0060 G99 G81 X7.75 Y5.3125 Z-0.25 R-7.25 F3.5 M03 center drilling of hole 1, 0.2-inch deep; (note that this program does not concern itself with the tool point-to-part surface distance or tool length offsets in order to concentrate on the canned cycle programming aspects)

N0070 X-0.4375 Y0.4375 center drilling of hole 2; (observe that the control only requires the location of this hole; the canned cycle remains active. G99 from N0060 will rapid return the tool point to R level from which the next hole will be machined following repositioning to the still active Z-level)

N0080 X0.4375 Y0.4375 hole 3

N0090 X0.4375 Y-0.4375 hole 4

N0100 X2.0625 Y1.5 hole 5

N0110 X0.75 hole 6

N0120 X0.75 hole 7

N0130 Y-0.75 hole 8

N0140 G98 Y-0.75 M05 hole 9; return to INITIAL; spindle stop

N0150 G80 M09 cancel canned cycle; stop coolant

N0160 G28 Z0 quill returns "home"

N0170 M06 T03 tool change, select tool No. 3

N0180 S1200 M03 select spindle speed, spindle start clockwise (note that in sequence N0050 we have only programmed the spindle speed, with the actual spindle start in the next sequence; this and other apparent discrepancies are intended to show that instructions may be written in a number of ways, for increased programming flexibility)

N0190 G99 G81 Z-0.425 R-6.06 F4.0 M08 drill hole 9; the G81 canned cycle was reset with the distances corresponding to the 0.25-inch drill; the tool change has allegedly taken place at the same location therefore no X and Y are programmed in N0190, and the drilling will take place in reverse order, thus avoiding unproductive machine travel

N0200 Y0.75 drill hole 8

```
N0210  Y0.75   drill hole 7
N0220  X-0.75   drill hole 6
N0230  G98  X-0.75  M09   drill hole 5; return to INITIAL level
N0240  G80  M05   cancel canned cycle; spindle stop
N0250  M06  T05   tool change; select tool in magazine holder No. 5
N0260  G99  G81  X-2.0625  Y-1.5  Z-0.65  R-6.875  F4.0   drill hole 4
N0270  X-0.4375  Y0.4375   drill hole 3
N0280  X-0.4375  Y-0.4375   drill hole 2
N0290  G98  X0.4375  Y-0.4375  M05   drill hole 1; return to INITIAL level
N0300  G80  M09   cancel canned cycle; coolant off
N0310  G28  X0  Y0   return to machine origin
N0320  M30   end of program.
```

The program could have been written in absolute, based on Fig. 9-2.

In comparison with conventional programming, the canned cycle operation is far more cost-effective. The above canned cycle program could be made even more efficient by the use of the subroutine feature, when several operations (center drill, drill, tap, etc.) are performed at the same locations. The location pattern is placed in a subprogram known as subroutine and called as needed by the main program which sets up the machine and changes tools. Any straight-line equally spaced hole pattern program can be drastically shortened by the use of the "L" word address, representing the number of repeats of the *incremental* displacement set up in the X-Y plane.

```
N0010 . . . .   sets up the canned cycle
N0020 G91 X . . . Y . . . L . . . the above canned cycle will be re-
      peated L times, each time at a position shifted over by X and Y from
      the preceding tool location, hence G91
```

The following application illustrates *repetitive canned cycle programming*. See Fig. 9-4.

The part drawing shows two straight-line equally spaced hole patterns. The upper row has 20 spaces, X-only motions. The overall distance being 10.0 inches, the spacing is 0.5 inch.

The lower row has 10 spaces. The X-space is provided by the same overall 10.0-inch distance as above, hence an X-spacing of 1.0 inch. The Y-space is obtained by dividing the 10 spaces into the 4.0-inch dimension, for a Y-spacing of 0.4 inches.

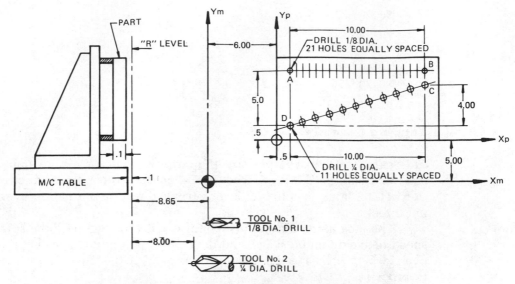

FIG. 9-4. Repetitive canned cycle programming.

In both cases, the canned cycle will be set up for the first hole in the pattern, while a single repetitive cycle block will take care of all the remainder.

N0010 G20 G40 G80 G91

N0020 B180

N0030 M06 T01

N0040 S2200 M03

N0050 G99 G81 X6.5 Y10.5 Z-0.25 R-8.65 F3.0 M08 This block will drill the first hole at point "A" in the upper row

N0060 G91 X0.5 L20 The sequence of operations initiated by N0050 will be repeated 20 times. G91 is redundant in block 0060 since it appears in block 0010, but it has been placed here to emphasize the incremental nature of "L." The reiteration of G91 will not create any problems, and the tool will travel from hole to hole at R level

N0070 G80 M09

N0080 G28 Z0 M05

N090 M06 T02

N0100 S1600 M03

N0110 G99 G81 Y-1.0 Z-0.25 R-8.0 F3.5 M08 This block will drill the first hole of the lower row at point "C." The remaining 10 holes will be drilled in a single repetition cycle block in the next sequence

N0120 G91 X-1.0 Y-0.4 L10 The sequence of operations initiated in sequence N0110 will be repeated 10 times at successive equally spaced locations

```
N0130 G80 M09
N0140 G28 Z0 M05
N0150 G28 X0 Y0
N0160 B180
N0170 M30
```

Tapping Canned Cycle G84

This canned cycle will perform the tapping of one or more holes as follows: Rapid to "R" level, feed to "Z-" level, reverse direction of rotation at "Z-" level, feed to "R" level, reverse direction of rotation to original direction.

The programmed feed must be fully synchronized with the spindle speed to avoid tool breakage, and the control will ignore speed and feed overrides, as well as feed hold and single block to the end of the sequence in progress.

To avoid an unproductive lengthening of the program, we'll assume that drilling has already taken place and that the "R" level is on the safe side of the clamps. Before starting the tapping portion of the program at, e.g., sequence N0100, we have to set tapping speeds and calculate tapping feeds.

Suppose the tapping speed has been selected to be 100 rpm (S100), the tapping feed for $3/8''$-16 could be calculated as follows:

1. Sixteen tpi represent a pitch of $1/16$ or 0.0625 inch. For one revolution of the tap, the quill will advance 0.0625 inch. For 100 revolutions, the quill advance or feed will be 100×0.0625 or 6.25 inches. As the above 100 revolutions take place in 1 minute (rpm), the amount 6.25 represents ipm and it is the tapping feed, F6.25.

2. It may be shorter to just divide the selected speed by the number of tpi. Therefore, $100/16 = 6.25$ and we program F6.25. Note that the programmer should use the maximum number of decimal places for maximum accuracy of the process.

3. When using an "inch" tap in a metric program, we just "soft-convert" the feed resulting from the method above to millimeters. F6.25 becomes F158.75, by multiplying 6.25 inches by the factor 25.4.

4. For metric threads in a metric program, since these are defined as the diameter times the pitch in mm, we just multiply the pitch by the selected rpm as in method 1. Using 125 rpm to tap M6x1, the feed will be 125×1 or F125.

5. For metric threads in an inch program (highly improbable, but not impossible), we simply divide the feed obtained as in method 4 by 25.4. Hence, 125/25.4 will be programmed as F4.921259, or on some recent controls as E4.921259.

The other tapped hole in our program is ⁵/₁₆″-18 tpi. The feed will be, given 100 rpm tapping speed: ¹⁰⁰/₁₈ = F5.5555556. For the last hole size, at 125 rpm, a ¹/₄″-20 will be tapped at F6.25.

N0100 S100 M03 spindle on, 100 rpm, clockwise. Coolant assumed on

N0110 G99 G91 G00 G84 X10.0 Y6.5 R-7.5 Z-0.5 F6.25 tool No. 1, assumed changed prior to seq. 0100 will rapid to the incremental dimensions X and Y, will rapid to "R," will tap to "Z," reverse and return to "R" where the clockwise rotation will be reestablished

N0120 G98 Y1.0 tap hole 2

N0130 G80

N0140 M05

Assuming that tool changing and offsetting for tool No. 2 has taken place according to the data in Fig. 9-1, we continue with sequence 180.

N0180 S100 M03

N0190 G99 G84 X-2.5 Y-0.5 R-7.25 Z-0.5 F5.556 tap hole 3

N0200 X-1.0 Y-1.0   tap hole 4

N0210 X-1.0 Y1.0   tap hole 5

N0220 G98 X1.0 Y1.0   tap hole 6, return to INITIAL LEVEL.

Following tool changing and offsetting for tool No. 3, we can reestablish the tapping canned cycle at hole 7, then program 10 repeats with an X pitch of 0.5 inch. Assuming sequence 260, we continue:

N0260 S125 M03

N0270 G99 G84 X-1.5 Y1.0 R-7.75 Z-0.5 F6.2S

N0280 X0.5 L10

Following completion of tapping, the program may be terminated the usual way or continued for other operations.

Fine Boring Canned Cycle G76

Fine boring canned cycle G76 code is one of a variety of boring cycles designed by the various manufacturers to cover just about any industrial application.

The canned cycle will perform the boring of one or more holes as follows: Rapid to "*R*" (point A, Fig. 9-5). Feed to "*Z*" (point *B*), spindle stop and orient, shift the tool point by the programmed amount *Q* away from the hole wall, rapid return to "*R*" or INITIAL as programmed, and restart of spindle. The following program will be in absolute, and as usual will place its emphasis on the canned cycle it illustrates.

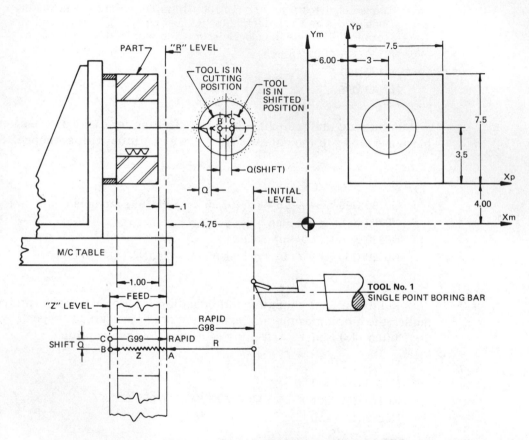

FIG. 9-5. Fine boring canned cycle G76.

N0010 G20 G40 G80 G90 G00
N0020 T01 M06
N0030 G92 X0 Y0 Z0   this command will zero the registers with
                       no motion taking place
N0040 S600 M03
N0050 M08
N0060 G98 G76 X9.0 Y7.5 Z-5.9 R-4.75 Q0.1 F0.75

At 600 rpm, F0.75 will ensure 800 tool marks or ridges in 1 inch of bored surface. Should a finer surface be required, we can increase S, reduce F, or both. In a production situation, a carbide-tipped boring tool could run at speeds to 3,000 rpm and feeds to 3.5 ipm. The introduction of fine boring cycles to CNC machining centers, particularly when the spindle is temperature-controlled, has drastically reduced the use of expensive conventional jig-borers. The above boring program can of course be continued or finished to suit as necessary.

Boring Canned Cycle G89

Somewhat similar to the precedent, this code will spotface, bore, or counterbore. The canned cycle will perform as follows: Rapid to "R," feed to "Z," dwell at "Z" for a duration given by "P" and rapid return to "R" or INITIAL depending on whether G99 or G98 was called in the program. See Fig. 9-6. One second is P100 and 3 seconds are P300, depending on the control, and decimal point programming does not usually apply to the P tape word. The program application below will illustrate both boring modes discussed in this chapter. Note that because of the central boss, the figure shows two "R" levels. See Fig. 9-7.

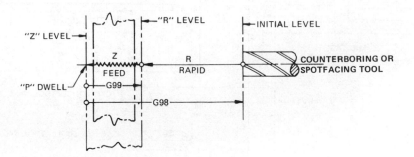

FIG. 9-6. Boring canned cycle G89.

We shall assume that the part has been supplied with the two 0.875 holes predrilled, but not bored, and the program will look as follows:

```
N0010 G20 G40 G80 G91
N0020 B180
N0030 T01 M06
N0040 S1600 M03
N0050 M08
N0060 G99 G81 X12.5 Y10.5 Z-0.25 R-7.5 F3.0   center drill hole 1
```

N0070 G98 X-6.0   center drill hole 2, return to INITIAL to avoid crash
on the way to hole 3

N0080 G99 Y7.5   center drill hole 3

N0090 X2.0 L3   center drill holes 4, 5, and 6

N0100 G28 Z0 M09   retract quill to machine Z-origin

N0110 G80 M05   cancel canned cycle, stop spindle

N0120 T02 M06

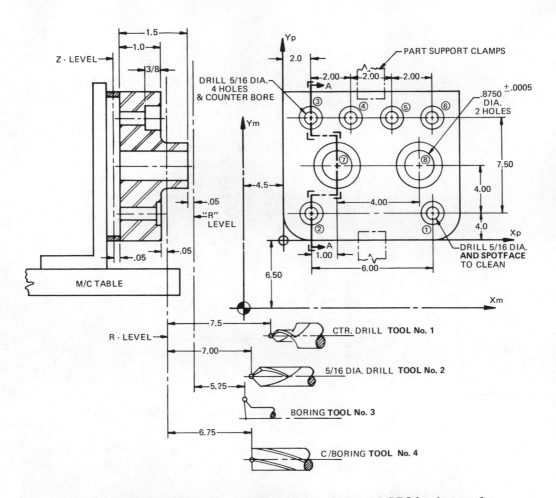

FIG. 9-7. Sample part illustrating G89 and G76 boring cycles.

If there is sufficient clearance, performing a tool change on top of hole 6
will allow immediate continuation of the machining process without idle
X- and Y-travel off the part and back on again.

```
N0130 S1500 M03
N0140 G99 G81 R-7.0 Z-1.2 F3.5      drill hole 6
N0150 X-2.0 L2                      drill holes 5 and 4
N0160 G98 X-2.0                     drill hole 3, return to INITIAL
                                       LEVEL
N0170 G99 Y-7.5                     drill hole 2
N0180 G98 X6.0                      drill hole 1
N0190 G80 M05                       cancel canned cycle, spindle
                                       stop

N0200 G28 Z0 M09
```

Again we have assumed that we can carry out a tool change on top of the part. As boring tools are usually longer, dimensional verification is advised.

```
N0210 T03 M06
N0220 S1500 M03
N0230 G99 G76 X-1.0 Y4.0 R-5.25 Z-1.6 Q0.1 F0.75   fine bore hole 8
N0240 G98 X-4.0   fine bore hole 7
N0250 G80 M05
N0260 G28 Z0 M09
N0270 T04 M06
N0280 S500 M03
N0290 M08
N0300 G99 G89 X-1.0 Y-4.0 R-6.75 Z-0.1 P100 F2.0   spotface hole 2
N0310 G98 X6.0   spotface hole 1, same depth as hole 2, then return to
      INITIAL for clearance
N0320 G99 Y7.5 R-6.75 Z-0.425   spotface hole 6 at 3/8-inch depth
N0330 X-2.0 L3   spotface holes 5, 4, and 3
```

The program can now be finished or continued as desired. Canned cycles may be initiated after a G80 or may just be changed from one canned cycle into the other "on the fly" without canceling in between. When a particular canned cycle is in effect, specific parameters may also be changed without requiring the reinitialization of the cycle. As in all other cases, thorough knowledge of the programming manual and the machine can extensively improve the efficiency of the program.

High-Speed Peck-Drilling Canned Cycle G73

Also known as woodpecker, intermittent feed, or deep hole drilling, the high-speed peck-drilling canned cycle is used to break the drill swarf in holes whose depth exceeds 2.5 times their diameter. See Fig. 9-8.

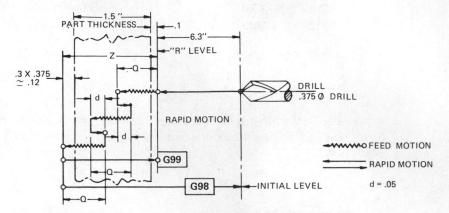

FIG. 9-8. High-speed peck-drilling canned cycle G73.

This canned cycle performs high-speed peck-drilling as follows: Rapid to *X-Y* location and to "R" level, feed by the amount "Q," rapid return by an unprogrammed fixed amount "d," feed and return alternately until "Z" is reached, and rapid return to "R" or INITIAL, depending on whether G99 or G98 has been programmed. The return amount "d" is usually set internally by control "parameters," and for our discussion it will be assumed to be 0.05. Our figure shows three "pecks" and two "interruptions" corresponding to three "Q"s and two "d"s, respectively. Since the total amount in feed, "Z" is shown as three times "Q" less twice "d," we can calculate "Q" by the following formula:

$$Q = \frac{\text{distance 'Z'' + (Number of interruptions} \times \text{``d'')}}{\text{Number of interruptions} + 1}$$

The distance "Z" represents the clearance above part + hole depth + 0.3 · drill diameter for clearance below the part. On some controls, one only needs to select a suitable "Q" value and the control will perform the necessary calculations and adjustments. In our figure,

$$Q = (1.72 + 2 \times 0.05)/3 = 0.61$$

and assuming the drill is in line with the required location, the tape block could read:

N0100 G91 G98 G73 R-6.3 Z-1.72 Q0.61 F3.0

Peck-Drilling Canned Cycle G83

The previous cycle was designed for self-removing swarf which had to be broken ever so often. Some materials, however, generate chips that will stick to the drill flutes, pack them, and score the hole before breaking the drill. The peck-drilling canned cycle G83 will rapid to X-Y location and to "R" level, feed by the amount "Q," rapid return to "R," rapid in by "Q-d," feed "Q" from there, rapid return to "R," thus removing the swarf from the part, rapid in by "$2Q$-$2d$," feed "Q" from there, rapid return to "R," etc. It can be seen that the total traveling distance is higher than that of the high-speed peck-drilling cycle, and must therefore be used only when required by the material-tool-coolant combination. See Fig. 9-9.

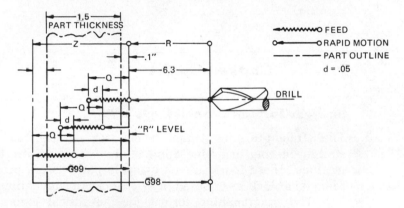

FIG. 9-9. Peck-drilling canned cycle G83.

The tape block is identical to the previous example, with the exception of the code itself:

N0100 G91 G98 G83 R-6.3 Z-1.72 Q0.61 F3.0

9.1.2   Turning Centers

Advanced computer technology has allowed the CNC system builders to incorporate additional canned cycles in almost every new model.

In the past few years, these changes have increased the number of canned cycle options from a few to a practically unlimited number. As a result, it is almost impossible to describe the programming or the cycle patterns of all existing systems within the limitations of a single textbook. We shall, however, discuss the systems most commonly used in the industry today. Learning new cycles should not present difficulties to anyone who understands the workings of those described in the following pages.

Under the topic of machining cycles for turning centers, we will discuss *canned cycles* and *multiple repetitive* cycles.

CANNED CYCLES: These cycles control four straight-line motions programmed in one tape block. The motions may be applied to turning, facing, or single-pass thread cutting. The first and last motions are in rapid, while the second and third are under controlled feed.

MULTIPLE REPETITIVE CYCLES: These cycles, on the other hand, are used for the programming of stock removal in external and internal turning applications, peck-drilling, multiple groove turning, and multiple thread turning.

9.1.3   Canned Cycles

Straight-Turning Canned Cycle G90

The straight-turning canned cycle can be used for a multitude of straight-turning or boring applications which require four straight-line motions. These four motions or cycle patterns are programmed in the same tape block, as opposed to four blocks in conventional programming.

The programming format for cylindrical external and internal turning is:

N__G90X__Z__F__

The G90 code, not to be confused with absolute programming in machining centers, defines the cycle pattern of the tool point as illustrated in Fig. 9-10.

In both internal and external mode, we program the X- and Z-dimensions of point 2 after bringing the tool point to the position marked 0.

To illustrate its use we will write a part program for the turning of a sample part illustrated by Fig. 9-11.

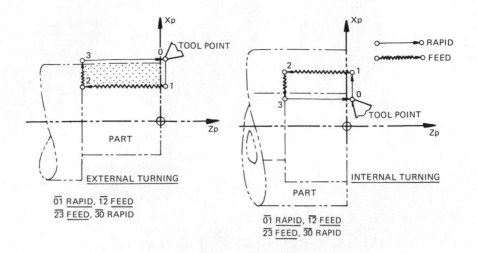

FIG. 9-10. Straight-turning canned cycle G90.

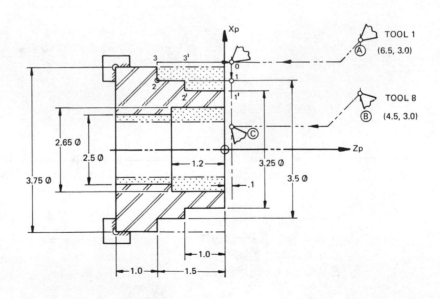

FIG. 9-11. Sample part illustrating G90 canned cycle.

N0010 G50 X0 Z0	Machine is at home position. Zero control
N0020 M41	Change gear to middle range
N0030 G98 T0101	Tool turret index to position 1, feed in ipm
N0040 S1200 M03	Spindle on, at 1,200 rpm
N0050 G00 X-8.0Z-18.5	Rapid to point A
N0060 G50 X6.5 Z3.0	Transfer coordinate system from machine to part
N0070 G00 X3.85 Z.1 M08	Rapid tool-point to "0"
N0080 G90 X3.5 Z-1.5 F7.5	Turn 3.5-inch diameter to 1.5-inch length

The G90 canned cycle guides the tool point through $\overline{01}$ (rapid), $\overline{12}$ (turn outside diameter in feed), $\overline{23}$ (face to 1.5 in feed) and $\overline{30}$ (rapid). The programmed X- and Z-dimensions are to point 2 in the part coordinate system. Without the G90 code, the tool point would follow an entirely different pattern.

N0090 X3.25 Z-1.0	Turn 3.25 diameter to 1.0 length

Since the G90 is modal we do not need to repeat it in subsequent tape blocks. The controlled motion pattern of block N0090 will be: $\overline{01'}$ (rapid), $\overline{1'2'}$ (feed), $\overline{2'3'}$ (feed) and $\overline{3'0}$ (rapid).

N0100 G00 X6.5 Z3.0	Return tool to point A

The G00 has a dual function in the tape block N0100. In addition to placing the control into rapid mode, it will also *cancel* the *canned cycle*.

N0110 G50 X-8.0 Z-18.5	
N0120 G30 U0.0 W0.0 T0808	Index tool turret to position 8
N0130 X-10.7 Z-16.0 S1450	Increase speed to 1,450 rpm
N0140 G50 X4.5 Z3.0	Reset the coordinate system to point B.

The preceding block will establish the tool point's dimensional relationship to the part coordinate system.

N0150 G00 X2.4 Z.1	Rapid tool point to "C"
N0160 G90 X2.5 Z-2.55 F6.0	Turn 2.5-diameter-bore
N0170 X2.65 Z-1.2	Turn 2.65-diameter bore to 1.20 depth
N0180 G00 X4.5 Z3.0	Rapid to point B, cancel G90
N0190 G50 X-6. Z-18.0 M05	Transfer the coordinate system from part to machine

The programmed X- and Z-dimensions are not necessarily the actual distance between the tool point and the machine home. However, if in the next tape block, using the G28 code, we program X- and Z-values in between the current tool location (B) and the programmed home (in block N0190) position, the control will drive the turret slide back to the machine home position. If the machine is equipped with a tailstock, the programmer should ascertain that the return motion will take the turret around it.

N0200 G28 X-2.0 Z-6.M09	Return the tool turret to machine home position

The above and other *return to home* blocks may be programmed only if at the beginning of the program (see block N0010) we have programmed a G50 X0 Z0 tape block to synchronize the control and the machine.

N0210 M30	End of program

NOTE: *If a tape block inside the canned cycle or following it contains only a sequence number and an end of block (EOB), the control will repeat the same cycle using the dimensions from the previous block.*

e.g.,

N0170 G90 X2.65 Z-1.2
N0180 E.O.B.

In N0180, N0170 will be repeated. *The G90 canned cycle must be canceled by a turret slide motion in G00 mode.*

Taper-Turning Cycle G90

The G90 canned cycle code can also be used to program tapered external and internal surfaces. Both the cycle pattern and the programming format must be changed to reflect the taper.

The programming format for tapered external and internal turning is:

$$N\_\_G90 \; X\_\_Z\_\_I\_\_F\_\_$$

The G90 controls the cycle pattern while the unit vector "I" defines the magnitude of the taper. The value of I may be positive ($+$) or negative ($-$), as illustrated in Fig. 9-12.

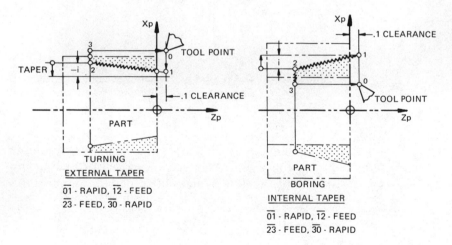

FIG. 9-12. Taper-turning canned cycle G90.

The unit vector "I" is programmed as the radius difference of the taper. The sign is viewed from the programmed point (2) to the feed start point (1). In either case, the clearance must be added to the part when we calculate the length of the tool pattern in feed.

A typical industrial application is shown in Fig. 9-13. To illustrate the use of the G90 code, we will write a part program for the turning of this part.

Fig. 9-13 shows the taper as a difference of two diameters over a given length, to which we add the tool clearance. Therefore, the change in diameter on the 2-inch length will be 0.60 inch, or 0.30 inch in terms of the radius. This radius difference is then divided into as many passes (we

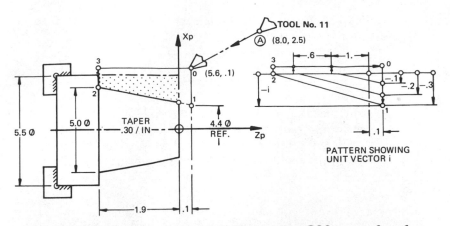

FIG. 9-13. Sample part (tapered) illustrating G90 canned cycle.

selected three) as the job may require. Once the cycle pattern and dimensions have been established, the program can be written as follows:

```
N0010 G50 X0 Z0
N0020 M41
N0030 G98 T1111
N0040 S900 M03
N0050 G00 X-10. Z-18.5
```

Programming this rapid traverse to point "A" requires the programmer to check the clearance between turret and tailstock to avoid a tool collision. Alternatively, this motion block can be replaced by two tape blocks, first by programming the Z, and second by programming the X-motion.

N0060 G50 X8.0 Z2.5	Transfer coordinate system from machine to part
N0070 G00 X5.6 Z0.1	Rapid to start point "0"
N0080 G90 X5.0 Z-1.0 I-0.1	Canned cycle start, first taper turning pass
N0090 Z-1.6 I-0.2	Second pass in canned cycle
N0100 Z-1.9 I-0.3	Finish pass in canned cycle
N0110 G00 X8.0 Z2.5	Canned cycle cancel and rapid tool point to "A"
N0120 G50 X-10.0 Z-18.5 M09	Transfer coordinate system from part to machine
N0130 G28 X0 Z0 M05	
N0140 M30	

NOTE: *Internal tapers are programmed identically, except for the sign of the unit vector "I"*

Facing Canned Cycle G94

Facing canned cycle G94 is identical to the straight-turning cycle so far as the cycle pattern is concerned. The difference is that the cycle motion starts with the "Z" instead of the "X." The first and last (fourth) motions are rapid and the second and third motions are in feed. The programming format is:

N__G94 X__Z__F__

The cycle can be applied to both internal and external facing operations. Typical patterns of this cycle are illustrated by Fig. 9-14.

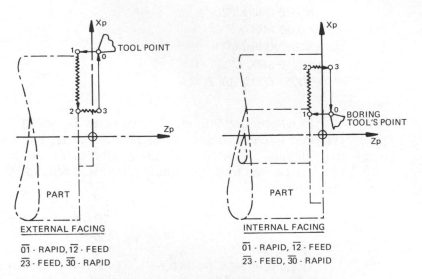

FIG. 9-14. Facing canned cycle G94.

In both external and internal facing, we program the X-Z coordinates of point 2. The sample part illustrated in Fig. 9-15 will outline the programming steps for the G94 code.

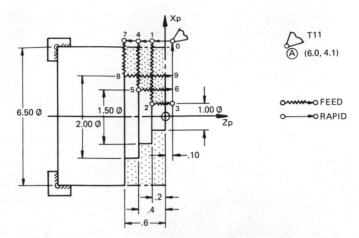

FIG. 9-15. Sample part illustrating G94 canned cycle.

N0010 G50 X0 Z0	Zero control to machine
N0020 M41	Second gear
N0030 G40 T1100 M08	Cancel offsets, index tool and turn coolant on
N0040 G00 X-6.0 Z-18.5	Rapid to point "A"
N0050 G97 S500 M03	Constant rpm, spindle on
N0060 G50 X6.0 Z4.1	Transfer coordinate system from machine to part
N0070 G00 X5.7 Z0.1 T1111	Rapid to point "0", tool offset register 11

The value in tool offset register 11 is zero at start; the operator can adjust for any dimensional inaccuracy during the turning operation without changing the program.

N0080 G96 S450	Constant surface speed
N0090 G94 X1.0 Z-0.2 F.01	Turn 1.00 diameter, 2 length
N0100 X1.5 Z-0.4	Turn 1.50 diameter, 0.4 length
N0110 X2.0 Z-0.6	Turn 2.00 diameter, 0.6 length

The G96 in block N0080 will vary the spindle rpm to maintain the programmed 450 fpm cutting speed. The G96 should be programmed when the change in the part diameter is larger than 1 inch. The motion pattern in blocks N0090, N0100, and N0110 can be described as:

N0090	$\overline{01}$—rapid, $\overline{12}$—feed, $\overline{23}$—feed and $\overline{30}$—rapid
N0100	$\overline{04}$—rapid, $\overline{45}$—feed, $\overline{56}$—feed and $\overline{60}$—rapid
N0110	$\overline{07}$—rapid, $\overline{78}$—feed, $\overline{89}$—feed and $\overline{90}$—rapid
N0120 G00 X8.2 Z4.1 M05	Rapid to point "A", cancel G94 canned cycle and stop spindle
N0130 G97 S500	Will change the constant surface speed to constant rpm
N0140 M30	End of program

This program has nine tape blocks less than an equivalent conventional part program without the G94 canned cycle. The program can be further improved by reducing the idle feed motion distances in blocks N0100 and N0110 from 0.5 inch ($\overline{56}$) and 0.7 inch ($\overline{89}$) to 0.3 inch and 0.5 inch, respectively. The change in the program can be implemented as follows:

N0090 G94 X1.0 Z-0.2 F.01	No change
N0100 G00 Z-0.1	Move canned cycle start point to Z-0.1-inch location
N0110 G94 X1.5 Z-0.4 F.01	Reprogram the canned cycle start point
N0120 G00 Z-0.3	Move canned cycle start point to Z-0.3-inch location
N0130 G94 X2.0 Z-0.6 F.01	Reprogram the canned cycle start point

The rest of the program remains unchanged. The saving in terms of machining time is insignificant in our example; however, it may be substantial in some other applications.

Taper Face Turning Canned Cycle G94

The programming is similar to the G90 taper turning canned cycle. The difference is the change of taper from the X- to the Z-axis, and the unit vector from I to K. The sign of the unit vector "K" is established with the same convention as we used for the cylindrical taper turning. The tape block format is written as:

$$N\_\_ G94\ X\_\_Z\_\_K\_\_F\_\_$$

The cycle patterns for the G94 canned cycle are illustrated in Fig. 9-16 for both internal and external face turning.

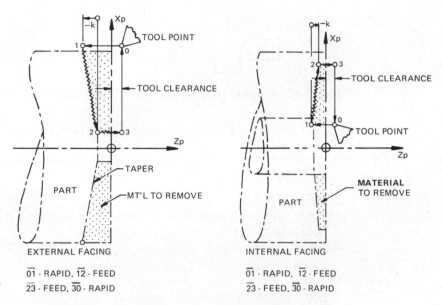

FIG. 9-16. Taper-face turning canned cycle G94.

The X- and Z-dimensions of point 2 are programmed in the canned cycle tape block. The direction of the unit vector "K" is viewed from this point to the start of the feed motion (point 1). If this direction (indicated by the arrows) is that of the (+) positive "Z-" motion, the sign of "K" will also be positive (+). Alternatively, the sign of "K" will be negative.

We will use the sample part shown in Fig. 9-17 to outline the programming steps for the taper face turning canned cycle, G94.

If the taper dimensions are readily available from the part drawing, as is the case in our example, we can proceed with the programming. However, if the taper is specified as: "Taper per inch of diameter" (0.02/1-inch-diameter), the programmer must first convert this information into programmable dimensions. The calculation in most cases will

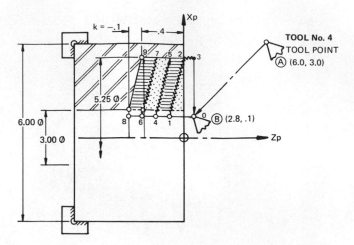

FIG. 9-17. Sample part (taper face) illustrating G94 canned cycle.

consist of simple additions or subtractions. This part program can now be written as:

N0010 G50 X0 Z0

N0020 M41 G99

N0030 G40 T0400 M08   Cancel offset, index tool, coolant on

N0040 G00 X-5.5 Z-16.5   Rapid to point "A"

N0050 G97 S900 M03   Constant rpm, spindle on

N0060 G50 X6.0 Z3.0   Transfer coordinate system from machine to part

N0070 G00 X2.8 Z0.1 T0404 Rapid to start point "B"; at start, register 04 is set to zero. Subsequently, the operator can adjust for dimensional inaccuracies without having to change the program.

N0080 G96 S380   Constant surface speed

N0090 G94 X5.249 Z0 K-0.1 F .01 First pass in G94

We have programmed 5.249 inches, instead of the part dimension of 5.25 inches. If in each subsequent tape block, we further reduce the diameter by the same amount, then, after the fourth pass we can program a finishing cut. The few thousands left on for the finish cut will reduce the cutting forces substantially, resulting in better accuracy and surface finish.

N0100 X5.248 Z-0.2   Second pass

N0110 X5.247 Z-0.3   Third pass

N0120 X5.246 Z-0.395   Final roughing pass

N0130 X5.25 Z-0.4   Finishing pass

This method of rough and finish turning with the same tool is most economical for small (under fifty parts) production runs. If the finish turning is done with another tool (in the same setup), the program should be written to leave 0.010 to 0.050 inch on diameters and 0.005 to 0.025 inch on faces.

N0140 G00 X6.0 Z3.0 M09	Cancel G94 canned cycle and rapid to point "A"
N0150 G97 S900	Constant rpm
N0160	End of program

Once the unit vector is programmed, as shown in tape block N0090, it need not be repeated in subsequent blocks.

NOTE: *TNR compensation can be programmed for both G90 and G94 canned cycles. It should be programmed in the block preceding the G90 or G94 canned cycle.*

Multiple Repetitive Cycles

These are more complex than canned cycles. It is in this area that the CNC systems vary the most. These features are developed for special applications and involve substantial software programming by the manufacturer of the control unit.

Stock Removal Cycle G71

Stock removal cycle G71 is used when several turning passes are required to remove stock. The result is a part outline bounded by straight lines, slopes, and circular arcs. The tape format is developed to adapt to any shape or form of turning or boring cycle pattern.

The tape format below applies to General Numerics controls. Other controls may use slightly different formats.

N__G71 P(ns) Q(nf) U(Δu) W(Δw) D(Δd) F__S__

The cycle pattern for this tape block is illustrated in Fig. 9-18.

P(ns)	"P" is the address of the tape block number where the cycle starts (ns). This number is normally the number following the tape block with the G71 code.

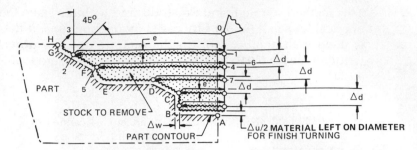

FIG. 9-18. Cycle pattern for stock removal cycle G71.

Q(fs)   "Q" is the address of the tape block number where the cycle ends (fs).

U(Δu)   "U" is the address of the material's dimension (Δu) left for finish turning on diameters.

W(Δw)   "W" is the address of the material's dimension (Δw) left for finish turning on faces.

D(Δd)   "D" is the address for the depth (Δd) cut. The Δd dimension cannot be programmed with decimal point. Example: 0.2-inch depth is D2000, 0.15-inch depth is D1500, 0.06-inch depth is D600

"F" and "S" are feed and speed word addresses.

The format within the cycle is restricted. Codes such as S, T, M, or G96, G97, may not be programmed between the starting sequence number, defined by "P" and the ending sequence number defined by "Q." These codes, if required, must be programmed prior to the G71 cycle block. The control will maintain these programmed codes for the validity of the multiple repetitive cycle. As shown in Fig. 9-18, "e" is the amount by which the tool point will retract from the part surface. This motion occurs at the end of each feed pass on a 45° angle. The cycle tape block sets up a subprogram to calculate the number of cuts or passes our tool must complete to produce the part contour. This contour is programmed in the tape blocks following the G71 code. Each segment or element is programmed by conventional means, using G01, G02, or G03, as the case may be, from point "A" to point "H." The cycle must be terminated by a G00, rapid motion block. The G71 code can be used in the same manner for internal contour turning.

For the pattern shown in Fig. 9-18, the tool point path can be described as follows:

First pass: Rapid to point 1, feed to point 2, feed to point 3 following the part contour, rapid by "e" at 45° toward start point, rapid return.

Second pass: Rapid to point 4, feed to point 5, feed to point 2 following the part contour and leaving the programmed (Δu) or (Δw) stock on for finish turning; rapid by "e" at 45° and rapid return.

The control will guide the tool point through the remaining passes, sometimes called "loops," until the contour is completed. Should the last loop be smaller than the Δd dimension, the control will automatically adjust the dimension.

To demonstrate the use of the G71 code, we will write a part program for the turning of a part illustrated in Fig. 9-19. The function of each block and code will be explained in detail.

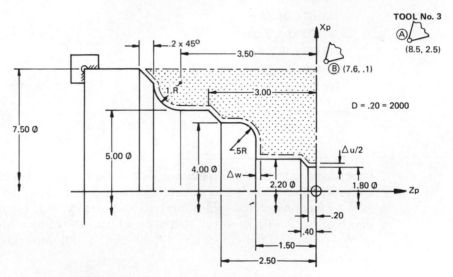

FIG. 9-19. Sample part illustrating the G71 cycle.

N0010 G50 X0Z0   Zero control to machine

N0020 G99 M41   Inches per revolution—ipr

N0030 G00 G40 T0300   Tool offset cancel, tool change to position 3

N0040 G00 X-6.8 Z-16.0   Rapid to point "A"

N0050 G50 X8.5 Z2.5 S1200   Transfer coordinate system from machine to part, set max. rpm at 1,200

N0060 G97 S800 M03   Start spindle at 800 rpm

N0070 G00 X7.6 Z0.1   Rapid to point "B"

N0080 G96 S300 M08   Constant surface speed; as the diameter of part decreases, the speed will be increased to maintain 300 fpm cutting speed

N0090 G71 P0100 Q0200 U0.04 W0.02 D2000 F0.012   This block initiates the stock removing cycle; the subprogram which describes the part contour starts in block 0100 (P0100) and ends in block 0200 (Q0200); 0.04/2 = 0.02 inch will be left on all the diameters (U0.04) and 0.02 inch on all the faces (W0.02). Depth of cut is set at 0.2 inch (D2000).

N0100 G00 X1.8
N0110 G01 F-0.2 F0.01
N0120 G01 X2.22 Z-0.4
N0130 Z-1.5
N0140 X3.0
N0150 G03 X4.0 Z-2.0 R0.5
N0160 G01 Z2.5
N0170 X5.0 Z-3.0
N0180 Z-3.5
N0190 G02 X7.0 Z-4.5 R1.0
N0200 G01 X7.5 Z-4.7

From sequence N0100 to sequence N0200 the program describes the part contour. Even though the first block N0100 defines a tool position of X = 1.80 diameter, the tool point will start the cycle at 0.2 inch (D2000) depth of cut. It will, however, define a part boundary into which the cutter or tool point will not enter. The "P" block (N0100) must only contain an X-motion.

N0210 G00 X8.5 Z2.5	Cycle cancel and return the tool point to point "A"
N0220 G97 S800 M09	Constant rpm
N0230 G50 X-6.8 Z-16.0	Transfer coordinate system from part to machine
N0240 G28 X0.0 Z0.0	Return turret slide to machine home position
N0250 M30	End of program

Internal contour turning can be programmed using the same format and programming steps.

Stock Removal in Facing G72

Stock removal in facing G72 is used when the repetitive turning motions are parallel with the X-axis. Accordingly, the "P" block must only contain a Z-motion. Examples of internal and external face turning using the G72 cycle are illustrated in Fig. 9-20.

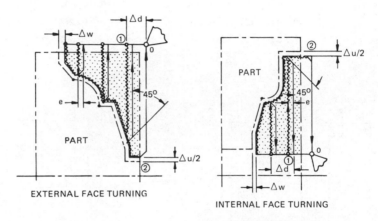

EXTERNAL FACE TURNING

INTERNAL FACE TURNING

FIG. 9-20. Cycle patterns for the G72 code.

The programming format is identical to the turning and boring cycle discussed previously, except for the depth of cut (d) which switches from the X- to the Z-direction.

The tool point path can be described as follows: Rapid to point 1 by the amount Δd, straight-line feed parallel to the X-axis, followed by feed along the part contour, rapid by amount "e" at 45° and rapid return to point 0.

The magnitude of the motion "e" is set by internal parameter. This cycle is then repeated by the control until the final depth is equal to or smaller than the "d" distance. After the final cut, the tool point will return to the start point 0. It should be noted that the tool point motion in feed in this particular feed cycle must either steadily increase in size or decrease from the start to the end. The control cannot perform this cycle if this requirement is not satisfied due to a profile containing a reverse contour.

The next part program will demonstrate the use of the G72 code. Figure 9-21 shows only the portion of the part directly related to the program.

External Turning
N0010 G50 X0.0 Z0.0   Zero control to machine

N0020 G99 M41	ipr, second gear of headstock
N0030 G00 G40 T0500	Tool offset cancel, tool change to turret position 5
N0040 G00 X-4.5 Z-20.	Rapid to point "A"
N0050 G50 X8.4 Z2.1 S990	Transfer coordinate system from machine to part and set maximum speed to 990 rpm
N0060 G97 S350 M03	Start spindle at 350 rpm
N0070 G00 X7.4 Z.1	Rapid to start point "B"
N0080 G96 S400 M08	Constant surface speed

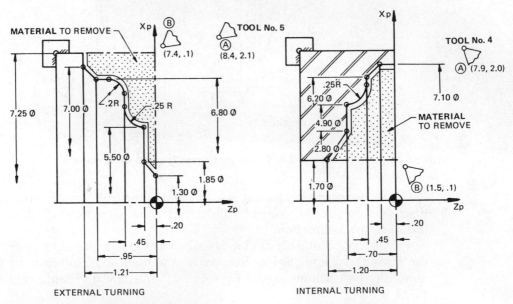

FIG. 9-21. Sample part illustrating the application of G72.

The change in diameter is 5.95 inches, therefore, it is necessary that a constant surface speed block be programmed. The constant surface speed not only will improve the program efficiency, but it will also maintain a constant cutting condition (load on tool, chip breaking, etc.).

N0090 G72 P0100 Q 0170 U.05 W.02 D1500 F.01

We shall now initiate the stock removing cycle G72 and set the limits for rough turning of the contour, as programmed in the following tape blocks:

N0100 G00 Z-1.21
N0110 G01 X 7.0 F0.007
N0120 X 6.8 Z-0.95
N0130 Z-0.65
N0140 G02 X6.4 Z-0.45 R0.2
N0150 G01 X 6.0
N0160 G03 X 5.5 Z-0.2 R0.25
N0170 G01 X1.85
N0180 X 1.3 Z0

Between sequences 0100 and 0180, the program describes the part boundary. The tool point will stay out of this boundary by the set $U/2$ and W dimensions programmed in block N0090. The first motion block inside the cycle or loop contains a Z-motion only.

N0190 G00 X 8.4 Z2.1	cancel G72 cycle and return the tool point to point "A" in rapid traverse
N0200 G97 S350 M09	change from constant surface speed to constant rpm
N0210 G50 X-4.5 Z-20.0	transfer coordinate system from part to machine
N0220 G28 X0.0Z0.0 M09	return turret slide to machine home position and turn coolant off
N0230 M30	end of program

You know, at this stage, how the G72 code controls the tool point during the facing cycle. The following Internal Turning program will not contain additional explanations.

Internal Turning
N0010 G50 X 0.0 Z0.0
N0020 G99 M41
N0030 G00 G40 T0400
N0040 G00 X-5.5 Z-16.5
N0050 G50 X 7.9 Z2.0 S1000
N0060 G97 S950 M03
N0070 G00 X1.5 Z0.1 M08
N0080 G96 S375
N0090 G72 P0100 Q0160 U-0.04 W0.015 D1000 F0.012

```
N0100 G00 Z-1.2909
N0110 G01 X2.8 Z-0.7 F0.006
N0120 X4.9
N0130 G03 X5.4 Z-0.45 R0.25
N0140 G01 X6.2
N0150 X7.1 Z-0.2
N0160 Z0.0
N0170 G00 X7.9 Z 2.0 M09
N0180 G97 S950
N0190 G50 X-5.5 Z-16.5
N0200 G28 X0.0 Z0.0 M05
N0210 M30
```

The preceding programming examples should provide sufficient exposure into repetitive cycles for rough turning. The basic principles and steps for other controls are either identical or quite similar. It is recommended that you study the tape format and cycle patterns provided in the system programming manual, prior to attempting to write a part program.

Finish Turning Cycle G70

The finish turning cycle G70 is also known as single-pass contouring cycle. It can only be programmed after a G71 and/or a G72 multiple turning cycle within the same part program. As discussed earlier, the finish turning can be done using the same tool as for roughing in small production runs. However, a different tool should be used for medium-to-large production runs. The G70, like the G71 and G72 cycles, can only be performed in memory mode of operation. The tape format is as follows:

N__G70 P(ns) Q(nf)

The P and Q words must be identical to the ones programmed for rough turning. The finish turning tape block does not allow programming of T, S, or F commands; therefore, the tool code and speed codes must be programmed after the end of the roughing and before the finishing tape block. The feed code, on the other hand, should be programmed inside the loop of the rough turning cycle. This feed will be used only during the finishing cycle. Should the programmed feed be too coarse for the finishing, the programmer can compensate by using a higher speed, which can be programmed prior to the finishing cycle block (feed in ipm).

To demonstrate the use of the G70 code, we will alter the programs written for the rough turning of the parts shown in Figs. 9-19 and 9-21. Instead of rewriting all the programs, we shall insert additional tape blocks as the application may require.

Finish Turning Of The Part Shown In Fig. 9-19

Tape blocks N0010 to N0220 require no changes

N0230 M01   Optional program stop in the "on" position will allow the operator to inspect the part; the operator can override the M01 in the "off" position

N0240 G97 S1100 M03   Start spindle at 1,100 rpm. When the M01 switch is turned off, the spindle speed will be increased to 1,100 rpm

N0250 G00 X7.6 Z0.1 M08   Rapid tool point to start position, coolant on

N0260 G96 S500   Constant surface speed has been increased from S300 (N0080) to S500. The increased surface speed will result in a higher (66%) rpm and higher quality surface finish even though the programmed feed rate of F0.01 has not been changed.

N0270 G70 P0100 Q0200   Initiates the finish turning cycle. The tool point will be guided along the part boundary derived from Seq. No. N0100 to N0200 inclusive

N0280 G97 S800   Constant rpm

N0290 G00 X8.5 Z2.5   Rapid the tool point back to point "A"

N0300 G50 X-6.8 Z-19.5 M09   Transfer coordinate system from part to machine

N0310 G28 X0.0 Z0.0   Return turret slide to machine home position

N0320 M30   End of program

Since the finish turning cycle is utilizing the tape block written for rough turning, it is mandatory that the first tape block (N0100) inside the loop be programmed in rapid motion. The second block (N0110) is for machining; therefore, it must be in linear (G01) or circular (G02 or G03) interpolation mode, combined with the appropriate feed code.

Finish Turning Of The Parts Shown in Fig. 9-21:

External Turning

Tape blocks N0010 to N0180 will not change.

N0190 M01

N0200 G97 S500 M03

N0210 G00 X7.4 Z0.1 M08

```
N0220 G96 S600
N0230 G70 P0100 Q0180
N0240 G97 S350 M09
N0250 G00 X8.4 Z2.1 M05
N0260 G50 X-4.5 Z-20.
N0270 G28 X0.0 Z0.0
N0280 M30
```

Internal Turning

Tape blocks N0010 to N0160 will not change.

```
N0170 M01
N0180 G97 S1150 M03
N0190 G00 X7.9 Z2.0 M08
N0200 G96 S525
N0210 G70 P0100 Q0160
N0220 G97 S950 M09
N0230 G00 X7.0 Z2.0 M05
N0240 G50 X-5.5 Z-16.5
N0250 G28 X0.0 Z0.0
N0260 M30
```

Pecking Cycle G74

The pecking cycle G74, a multiple repetitive cycle pattern, is identical to the peck-drilling canned cycle discussed earlier. The repetitive motion of the tool turret can be used for either turning grooves in the part face or drilling a hole in the center of the part. In both cases, the cutting process requires an interruption of the feed motion to break up the continuity of the chip. The pecking-cycle standard on most turning centers for drilling is an optional feature for grooving.

Multiple Grooving Cycle G74

This cycle is used for the turning of deep multiple grooves. Since the turning cycle controls the Z-axis in feed and the X-axis in rapid motion, we can turn any number of grooves using only one tape block.

If we assume a feed motion of $K = 0.2$ inch, it would take a minimum of 18-tape blocks in conventional programming. Instead, we can write one single tape block as follows:

$$N\_\_ G74 \, X\_\_Z\_\_I\_\_K\_\_F\_\_$$

using the G74 cycle.

The individual addresses are marked on the cycle pattern drawing shown in Fig. 9-22.

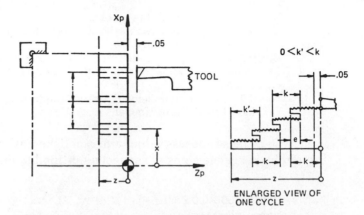

FIG. 9-22. Cycle pattern for peck-grooving cycle G74.

This cycle allows turning equally spaced grooves. The "e" rapid retract dimension is set by an internal parameter.

To demonstrate the use of the G74 peck-grooving cycle, we shall write a part program for the turning of the sample part illustrated in Fig. 9-23.

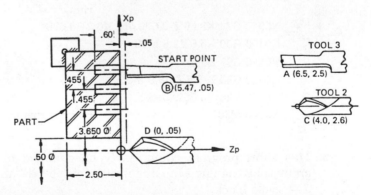

FIG. 9-23. Sample part illustrating the G74 cycle.

The dimensions for each address of the G74 cycle are established as follows:

X = 3.650 diameter, diameter of the last groove turned in the cycle. The turning sequence may be reversed to start the groove at 3.65

diameter; however, the X-dimension would have to be changed to 5.47 diameter. This value was calculated as follows:

$$3.65 + 2(0.455 + 0.455) = 5.47$$

$Z = -0.6$, the depth of the groove

$I = 0.455$, distance between the grooves; this distance cannot vary from groove to groove.

$K = 0.2$, distance of Z-motion in feed between interruptions. K is selected by the programmer. Both I and K should be programmed unsigned.

$F = 0.006$, turret slide velocity during the feed motion. F is also selected by the programmer.

Once these address values are established, we can write the part program by using the steps of sequence outlined below:

N0010 G50 X0.0 Z0.0

N0020 G99 M41

N0030 G00 G40 T0300

N0040 G00 X-8.5 Z-18.5   Rapid tool to point "A"

N0050 G50 X6.5 Z2.5 Z1200   Transfer coordinate system from machine to part and set maximum spindle rpm

N0060 G97 S650 M03

N0070 G00 X5.47 Z0.05 M08   Rapid tool point to point "A"

N0080 G96 S450   Set constant surface speed at 450 rpm

N0090 G74 X3.65 Z-0.6 I0.455 K0.2 F0.006

N0100 G00 X6.5 Z2.5 M09

N0110 G97 S650

N0120 G50 X-8.5 Z-18.5 M05

N0130 G28 X0.0 Z0.0

N0140 M30   End of program

This short program could easily be changed to turn any number of grooves by changing the dimensions of the "X" and "I" addresses in the G74 cycle block.

Peck-Drilling Cycle G74

The peck-drilling cycle G74 is used for drilling deep holes. The tape format is different from the peck grooving. Since the hole can only be drilled in the center of the part, there is no X-motion in the cycle. The subroutine, albeit using the same G74 code, can distinguish between the drilling and the turning cycles because of the change in the format.

N__ G74 Z__ K__ F__

We shall write a part program for the drilling of the 0.50-diameter hole in the center of the part shown in Fig. 9-23.

```
N0010 G50 X0.0 Z0.0
N0020 G99 M41
N0030 G00 G40 T0202
N0040 G00 X-8.2 Z-16.
N0050 G50 X4.0 Z2.6 M08
N0060 S850 M03
N0070 G00 X0.0 Z.05
N0080 G74 Z-2.65 K0.5 F3.5
```

The G74 cycle will start the peck drilling to a total Z-depth of -2.65, including the tool point angle. The cycle will repeat the interruption at every 0.5 ($K = 0.5$) distance until the total depth is drilled, then the tool point will return in rapid motion to the starting point "D."

```
N0090 G00 X4.0 Z2.6 M09
N0100 G50 X-8.2 Z-16.0 M05
N0110 M30                          End of program
```

This cycle bears no resemblance to the peck-drilling cycles discussed under the machining centers. This cycle was developed strictly for the drilling of one single hole per part.

Multiple Groove Turning Cycle G75 (In the X-axis Direction)

The multiple groove turning cycle G75 pattern is identical to G74 if the X- and Z-axes, as well as the "I" and "K" addresses, are reversed. The two programming formats will therefore be very similar. The function of the individual addresses can be described as follows:

X = bottom diameter of the grooves; this must be the same for all the grooves turned in the cycle block

Z = dimension from Z-part zero to the last groove in the cycle

I = is the depth of the tool point motion in feed between interruptions

K = is the dimension between the equally spaced grooves in the cycle

The use of the G75 groove turning cycle will be shown in Fig. 9-24.

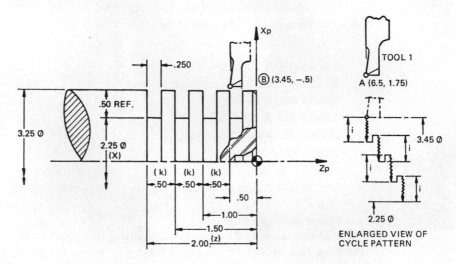

FIG. 9-24. Sample part illustrating the G75 cycle.

The program may be written with the starting groove nearest to the chuck or the tailstock by changing the Z-dimension in the block preceding the G75 cycle block.

N0010 G50 X0.0 Z0.0

N0020 G99 M41

N0030 G00 G40 T0100

N0040 G00 X-6.0 Z-18.5

N0050 G50 X6.5 Z1.75 S1200

N0060 G97 S850 M03

N0070 G00 X3.45 Z-0.5 M08

N0080 G96 S400   The constant surface speed will produce constant cutting conditions. As a result, even surface finish.

N0090 G75 X2.25 Z-2.0 I.2 K.5 F.005   As in the previous example, the "I" and "K" values are programmed unsigned.

N0100 G00 X6.5 Z1.75

N0110 G50 X-6.0 Z-18.5

N0120 G97 S850 M09
N0130 G28 X0.0 Z0.0 M05
N0140 M30   End of program

Thread Cutting

CNC system builders have developed specific canned cycles for the turning of straight, tapered, and scroll threads for single, double, and multiple starting threads. Most current CNC turning centers are equipped with straight- and taper-threading cycles for both internal and external thread turning as standard features. More complex cycles are offered on an optional basis at extra cost. In the following pages, we shall attempt to clarify in some detail the cycle patterns, tape formats, and the programming of thread turning. Because of their similarities, the threading cycles are presented as a separate group.

Single Start Multipass-Thread-Turning Cycle G92

The single start multipass-thread-turning cycle G92 is the simplest canned cycle for the turning of straight and tapered threads for internal and external applications. The cycle pattern is identical to the G90 turning and boring cycle from which it was derived. While the speed-feed ratio had no significance during the G90 cycle, in *threading, the speed-feed ratio has prime importance.* As we have earlier discussed, the lead of the thread is a function of the spindle rpm. If the tool point moves at 1 ipm, and the spindle is rotating at 100 rpm, the lead (pitch) of the thread will be 0.01 ipr. To produce a constant lead from the beginning to the end of the thread, the spindle must maintain a constant rpm. For this reason, all *the thread-turning cycles lock the programmed spindle speed and turret slide feed* during the cycle. Any adjustment of the speed or feed override circuits is ignored by the control. The programmer is advised to study the appropriate section of the programmer's manual for restrictions such as maximum rpm, minimum and maximum programmable lead, etc. These programmable parameters vary from system to system even for the same make of machine and control.

The 4NE turning center illustrated in the following discussion allows the programmer to use conventional feed programming to an accuracy of four decimal places. In this mode, the range of the feed address can vary from a minimum of F0.0001 to a maximum of F50.0000 ipm. Turning a part with 18 tpi would require $1/18 = 0.055556$ ipr feed rate. Since this feed programming format will allow us to program F0.0555 feedrate only, for longer threads, the inaccuracy can add up to several thousands of an inch. To compensate for this inaccuracy, we have the option of using an "E" instead of an "F" address. In the "E" mode, we can program our feed to six-decimal-point accuracy. The range of the "E" address can vary from E0.000001 to E9.999999 ipr providing a more flexible method for turning threads to gage accuracies.

The *maximum* programmable spindle speed (n) must be smaller or equal to a ratio of 196.85 (constant) divided by the lead (L), expressed in mathematical terms as:

$$n = \frac{196.85}{L}$$

For 12 tpi, the lead (pitch) $L = \frac{1}{12} = 0.0833333$ and the maximum rpm (n) must be less than or equal to:

$$n = \frac{196.85}{0.0833333} = 2,362 \text{ rpm}$$

Obviously this is far too high. Taking another example of 6 tpi, twice the 12 tpi:

$$n = \frac{196.85}{0.166666} = 1,181 \text{ rpm}$$

This is still a very high rpm for thread turning. See Fig. 9-25.

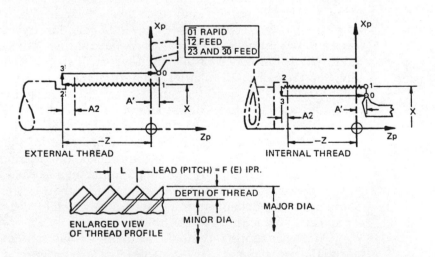

FIG. 9-25. Cycle pattern for the straight-thread-turning cycle G92.

The programming format is as follows:

N__G92 X__Z__F(or E)__

where

       N is the sequence number
       X is the tool point dimension at start point
       F is the feed in ipr to four decimal places accuracy or
       E is the feed in ipr to six decimal places accuracy.

Only one "F" or "E," but not both, can be used in the cycle block.

The cycle patterns for the G92 straight-thread-turning cycle are shown in Fig. 9-25.

The tool point must be at point 0 and the spindle in rotation prior to programming the G92 thread-turning cycle. Before discussing a sample program, we must look at the effect of the tool's acceleration A1 and deceleration A2 at the beginning and end of the thread. The dimension of A1 is the distance required by the tool point to reach the programmed feed rate. A2 is the distance required by the tool point to come to a stop from the programmed feed rate. These dimensions can either be calculated by a formula, provided in the programming manual, or taken from a chart or graph. For the 4NE turning center, the A1 and A2 dimensions can be calculated using the formulas below:

Spindle rpm	A1	A2
100	0.41 × Lead	0.2 × Lead
200	0.82 × L	0.4 × L
300	1.23 × L	0.6 × L
400	1.64 × L	0.8 × L
500	2.05 × L	1.00 × L
600	2.46 × L	1.15 × L
700	2.86 × L	1.4 × L
800	3.27 × L	1.6 × L
900	3.68 × L	1.8 × L
1,000	4.09 × L	2.0 × L
1,100	4.50 × L	2.2 × L
1,200	4.91 × L	2.4 × L

Turning an 8 tpi part at 100 rpm would require an acceleration distance of:

$$L = {}^1\!/_8 = 0.125$$
$$A1 = 0.41 \times 0.125 = 0.0512 \text{ inch}$$

and a deceleration distance of:

$$A2 = 0.2 \times 0.125 = 0.0250 \text{ inch}$$

However, turning the same thread at 1,200 rpm would require distances of:

$$A1 = 4.91 \times 0.125 = 0.6137 \text{ inch}$$

and

$$A2 = 2.4 \times 0.125 = 0.3 \text{ inch}$$

The programmer must therefore select the rpm to suit the clearances available on the part. The range is normally adequate to turn the thread to the designed tolerances.

Tapered Thread-Turning Using G92

The programmer must know the maximum angle (α) of the taper which can be turned on a particular CNC lathe. For the 4NE turning center, the maximum angle (α) is 45°.

The tape format for the programming of the G92 code for tapered thread is as follows:

N__G92 X__Z__I__F(or E)__

The cycle pattern is shown in Fig. 9-26.

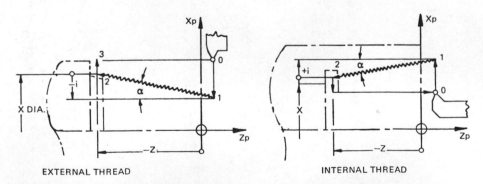

EXTERNAL THREAD   INTERNAL THREAD

FIG. 9-26. Cycle pattern for the taper-thread-turning cycle G92.

To demonstrate the use of the G92 code, we will write a part program for threads turning on the part shown in Fig. 9-27.

It will be assumed that the turning portion has already been completed. The following steps are required:

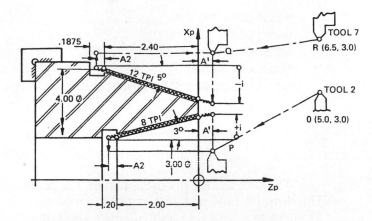

FIG. 9-27. Sample part illustrating the G92 taper-thread-turning cycle.

1. Calculate the lead $F(E)$ for both tools:
   Tool No. 2   8 tpi = $^1/_8$ = 0.125
   Tool No. 7   12 tpi = $^1/_{12}$ = 0.803333

2. Calculate the distances A1 and A2 required for the tool point acceleration and deceleration:

Selecting 400 rpm for tool No. 2 and 300 rpm for tool No. 7, the distances required can be worked out.

Tool No. 2   A1 = 1.64 × 0.125 = 0.205; we round it off to 0.21
              A2 = 0.8 × 0.125 = 0.1 inch
Tool No. 7   A1 = 1.23 × 0.083333 = 0.1025; round off to 0.11
              A2 = 0.6 × 0.083333 = 0.05

At these selected spindle speeds, both tools will have enough clearance to decelerate, therefore, our values will not affect the accuracy of the threads.

3. Establish the start point dimensions:
   for tool No. 7:
   Q (4.5, 0.11)
   for tool No. 2:
   P (2.5, 0.21)
   These values will allow adequate clearances at the end of the feed stroke for both tools to clear the part.

Having established the above three sets of dimensions, we can now proceed with the part program

```
N0010 G50 X0 Z0
N0020 M40 T0202 G99   tool No. 2
```

```
N0030 G97 S400 M03
N0040 G00 X-5.4 Z-15.
N0050 G50 X5.0 Z3.0 M08
N0060 G00 X2.5 Z0.21                    rapid to point "P"
N0070 G92 X2.8728 Z-2.1 I0.1208 F0.125  first pass
```

$$Z = -2.0 + (-A2) = -2.0 - 0.1 = -2.1$$
$$I = (2 + 0.1 + 0.205) \cdot \tan 3° = 0.1208$$

The G92 has set the cycle pattern for the multipass thread turning. The thread turned by tool No. 2 is a class 3 thread. Its major diameter is 3.000 inches, and its minor diameter is 2.8723 inches. From practical experience, six passes are required. As the depth increases, we will progressively reduce the depth of cut. The final pass for finishing will be repeated at the same depth to size the thread; accordingly, the cutting force in the finish pass will be approximately 5% of the preceding cutting pass, without any tool deflection. This will produce a part with uniform surface finish and accurate tolerance.

```
N0080 X2.9128        second pass
N0090 X2.9428        third pass
N0100 X2.9728        fourth pass
N0110 X2.9895        fifth pass, finished minor diameter
N0120 X2.9895        final pass
```

The programmed values of the Z, I, and F addresses will remain active and be repeated in each pass from N0080 to N0120.

```
N0130 G00 X5. Z3. T0200              cancel G92 cycle
N0140 S300 T0707                     tool No. 7
N0150 G50 X6.5 Z3.0
N0160 G00 X4.5 Z0.11                 rapid to Point "Q"
N0170 G92 X3.9687 Z-2.45 I-0.2233 E0.083333  first pass
```

The major diameter of the 4-12 tpi thread = 4.00, and the minor diameter = 3.8917. We will use again six passes for the thread turning.

$$Z = -2.4 + (-A2) + (-A1) = -2.4 - 0.05 - 0.1025 = -2.5525$$
$$I = 2.5525 \cdot \tan 5° = -0.2233$$

For accuracy, we have replaced the *F* address by *E*.

N0180 X3.9387	second pass
N0190 X3.9187	third pass
N0200 X3.9087	fourth pass
N0210 X3.9009	fifth pass, finished minor diameter
N0220 X3.9009	final pass
N0230 G00 X6.5 Z3.0 M09 T0700	cancel G92 cycle
N0240 G50 X-5.4 Z-15.0 M05	
N0250 G28 X0 Z0	
N0260 M30	

The latest CNC turning centers now offer an even more compact thread-turning cycle. This cycle allows the programmer to specify all the data required for any thread turning in one tape block.

Multipass Thread-Turning Cycle G76

The multipass thread-turning cycle G76 allows the programmer to define any number of passes in one tape block. The CNC system will perform straight and tapered, internal and external threading. The guidelines for the spindle rpm and tape calculations, outlined for the G92 cycle, do not change. The tape format is as follows:

N__G76 X__Z__I__K__D__F(E)__A__

where G76 is the multipass thread cycle,

X	is the minor diameter of the thread,
Z	is the length of thread plus the deceleration distance.
I	is the radius difference measured between the start and end points of the thread
K	is the depth of the thread, calculated as $K = \dfrac{D \text{ major} - D \text{ minor}}{2}$
D	is the depth of cut for the first pass (measured in terms of the radius, no decimal point)
F(E)	is the lead of the thread
A	is the angle of thread, normally 30°, 55° or 60°. The 4NE has additional 0°, 29°, and 80° programmable angles.

Since the cycle pattern is identical to the G92 code, we will not repeat the explanations. Instead, we shall write a part program for the thread turning of the same part. Using the same thread and tool data, we can proceed with the part program as:

```
N0010 G50 X0 Z0
N0020 M40 T0202
N0030 G97 S400 M03
N0040 G00 X-5.4 Z-15.0
N0050 G50 X5.0 Z3.0 M08
N0060 G00 X2.5 Z0.21
```

These tape blocks are identical to our previous program.

```
N0070 G76 X2.9895 Z-2.1 I0.1208 K0.0811 D150 F0.125 A60
```

This G76 cycle block replaced the tape blocks N0070 to N0120 and will produce the same 3-8 tpi thread. The dimensions used are identical but programmed in different formats. "K" is calculated as follows:

$$K = \frac{2.9895 - 2.8273}{2} = 0.0811$$

The thread profile angle "A" is defined by the part drawing.

```
N0080 G00 X5.0 Z3.0 T0200
N0090 S300 T0707
N0100 G50 X6.5 Z3.0
N0110 G00 X4.5 Z0.11
N0120 G76 X3.9009 Z-2.45 I-0.2233 K0.0540 D100 E0.083333 A60
```

The value of "K" is derived as follows:

$$K = \frac{4.0088 - 3.9009}{2} = 0.0540$$

```
N0130 G00 X6.5 Z3.0 T0700 M09
N0140 G50 X-5.4 Z-16.5 M05
N0150 G28 X0.0 Z0.0
N0160 M30                          end of program
```

We have reduced the program from 26 to 16 tape blocks. It would also require proportionally less memory core for the storing of this program.

Double-Start Thread-Turning

Either G92 or G76 codes can be used to turn a double start thread or, for that matter, any number of starts. The method is to program the cycle as in our previous examples, cancel the G92 or G76 code by programming a G00 move with a Z-motion. The Z-motion will be one-half of the "F" or "E" (lead) in the positive direction for a double start, one-third of the "F" or "E" for a triple start, and $1/n$ for "n" starts of thread.

Then repeat the thread turning cycle with the same dimensions in the following tape block.

If the thread has three starts, the thread-turning cycle has to be repeated three times. For "n" starts, the cycle will be repeated "n" times in the same program. Prior to repeating the cycle there must always be a rapid (G00) slide motion to cancel the previous cycle. This same tape block will also establish the start point for the next threading cycle.

Summary

Understanding of the canned cycles in this chapter will make it possible to write efficient part programs for the machining of complex parts. The programming steps and cycle patterns of the following cycles were discussed can be summarized as follows:

Canned Cycles for Machining Centers

These are preprogrammed subroutines permanently stored in the CNC system's memory. Every canned cycle has a fixed pattern and tape format. Each canned cycle is activated by a specific G-code.

G80 CANNED CYCLE CANCEL: When programmed, it will cancel a previously used canned cycle in the same or previous part program.

G81 DRILL CANNED CYCLE: When used in a part program, it will rapid the tool point to an "R" level, feed to a "Z-" depth and will return the tool point, in rapid motion, to the "R" level if a G99 code is programmed in the same tape block. Any number of holes may be drilled by programming the X-Y coordinates of the holes in the blocks following the G81 cycle. Equally spaced holes can also be drilled if the tape block contains the number of holes to be drilled in the "L" address.

G84 TAPPING CANNED CYCLE: The subroutine combines the necessary tool motions, linear Z-axis and rotary spindle, required for the tapping of predrilled holes. The programmer must calculate the correct feed and speed values for a specific thread in terms of the linear displacement corresponding to each spindle revolution. The tool rotates, rapids to "R" level, feeds to "Z-" level, reverses the spindle rotation and feeds back to "R" level, reverses the spindle to its original direction of rotation. The "L" address can also be programmed for equally spaced hole tapping to the same depth.

G73 HIGH-SPEED PECK-DRILLING CYCLE and G83 PECK-DRILLING CYCLE: These are used for deep hole drilling. The cycle can be programmed to interrupt the feed motion to allow the swarf to clear any number of times. The cycle pattern is identical to the G81 with the exception of the feed motion.

G76 FINE BORING CYCLE: This cycle is used for finish boring holes, with single point tools, to jig-bore accuracies. In the cycle pattern, the tool point will rapid to "R" level, feed to Z-depth; the spindle will stop, orienting the tool point; the tool will shift by an amount "Q" (in direction opposite to the tool point) and will rapid to "R" or initial level.

G89 BORING CYCLE: This cycle is used for spotfacing or counter boring applications.

Canned Cycles for Turning Centers

These are divided into two groups: *The canned cycles* normally control the tool point motions in a defined pattern for one cycle. The *Multiple Repetitive Cycles* will control the tool point through a number of cycles; while the shape of the pattern is maintained, the size of the motions is changing.

G90 TURNING-BORING CYCLE: This cycle is used for stock removing in both external and internal applications.

G90 TAPER TURNING CYCLE: This cycle is used for rough turning of internal or external tapers. The tape format must contain the radial taper information in the "I" address. The sign of "I", depending on the taper, may be positive or negative.

G94 FACING CYCLE: Like the G90 in turning, the G94 controls the tool point in facing applications. The magnitude of the taper is programmed under the "K" address; its sign may also be positive or negative. If the tape block with the G94 code has no "K" address, the format is for straight internal or external turning.

G71 Stock Removing Cycle: This cycle is used for rough turning of diameters with complex contours in several passes. The tape format defines the finished part's contour with allowances for finish turning. The contour, often called part boundary, may be defined through linear and circular interpolation. This cycle should be used in constant surface speed (G96) mode.

G72 Stock Removal in Facing: This cycle is identical to the G71 code, but it is used for facing. The G72 is used when several facing passes are required to remove a large stock of material.

G70 Finish Turning Cycle: This cycle is normally used after the G71 or G72 cycles, within the same program. The part contour is defined in the stock removing cycle; therefore, the G71 code sends the control back to repeat tape blocks without the allowances left on diameters and faces.

G74 Multiple Grooving and Peck-Drilling Cycle: Using the same tape format, the G74 code can be utilized to control peck-drilling as well as turning grooves in the face of a part.

G75 Multiple Groove Turning Cycle: The G75 code controls the X-turret slide motions in feed and the Z-turret slide motions in rapid mode.

G92 Single Start Multipass Thread-Turning Cycle: This cycle is used for turning straight and tapered internal and external threads.

G76 Multipass Multistart Thread-Turning Cycle: This cycle is used to program straight and tapered external or internal threads in one single tape block. Multistart threads can be turned by moving the start point of the turning tool by one-half lead distance for double start, one-third lead distance for triple start, and $1/n$ lead distance for "n" start threads.

9.2 VARIABLE CANNED CYCLE PROGRAMMING

As expected, there are slight variations in programming from control to control, but the principles are similar. Let us look at an example on a DoALL vertical CNC mill, with a DEC-PDP 11 based Elex 450 control.

Variable Canned Cycle for a Bolt Circle

Establishing the Variables

1. Number of holes
2. Radius of bolt circle (polar coordinates: radius and angle)
3. Starting point (hole) expressed by its radius and angle
4. Angular pitch (hole-to-hole angle)
5. A counter

Defining the Variables

This control uses the G59 code, followed by a digit, or a letter, enclosed in square brackets. The digits could be 0 to 9. The letters may not include the standard word addresses, such as N, G, X, Y, Z, I, J, K, T or M, if used in the same program.

e.g., Sequence 10, below, states that there are 16 holes.

N10,G59,[1] = 16

This information will be stored in a "storage register." The register's number is [1] and its content may change, or vary, hence the word variable.

Conditional Testing

As the control calculates the location, positions the tool, and performs the operation, it becomes necessary to be able to verify whether a certain value, such as the final number of holes has been attained. This verification is expressed as follows:

G53,[1] < 16   contents of register 1 less than 16
G54,[1] = 16   contents of register 1 equal to 16
G55,[1] > 16   contents of register 1 greater than 16

and may be used as needed, one at a time.

Setting the Counter

In sequence 40, below, a counting register, labeled D, will have its contents increased by one.

    N40,G56,D1

In sequence 50, the contents of counting register D will be compared with the required number of holes, 16. This number has been stored in register No. 1 during sequence 10.

    N50,G54,D[1]

So long as the contents of the counter (which counts the number of holes drilled) are less than the contents of storage register No. 1 (which contains the required number of holes), drilling will be continued.

In the following, we shall examine a bolt circle variable canned cycle. The data are:

- Number of holes: 16
- Bolt circle diameter: 3.75 inches (radius 1.875 inches)
- Depth of drilling: 1 inch
- Tool point "home": 0.2 inch above part surface.

N10,G59,[1] = 16   Register No. 1 contains the number of holes, 16

N20,G59,[2] = 1.875   Register No. 2 contains the bolt circle radius

N30,G59,[3] = 0   Register No. 3 contains the starting angle, 0°. Polar coordinates of hole 1 are (1.875,0), cartesian coordinates: X1.875 Y0

N40,G59,[4] = 360/[1]   The pitch angle will be 22.5°, calculated by the minicomputer in the CNC by dividing 360° by the contents of register No. 1

N50,G59,[5] = [3]   The contents of counting register No. 5 are set to the contents of register No. 3. Register No. 5 will act as the current angle counter.

N60,F45,G90,G11,R[2],C[5],Z0   At a feed of 45 ipm, in absolute mode G90, perform a linear interpolation G11 in polar coordinates to the location: Radius per register 2, i.e., 1.875 inches and angle C zero, per contents of register No. 5 which are equal to the contents of register No. 3. No Z-motion at this time.

N70,G91,F4.5,G01,Z-1.2   Incremental linear interpolation

N80,F45,Z1.2 Fast tool withdrawal to the initial position, at 0.2 inch above the part surface. Alternatively we could have written N80,G00,Z1.2

At this point in time, the first hole has been drilled.

N90,G59,[5] = [5] + [4] The contents of the counting register are now increased by the contents of register No. 4, i.e., the pitch angle 22.5°.

N100,G56,D1 An empty register, D, is selected, and its content, zero, is increased by one.

N110,G54,D[1] The contents of D, presently one, are compared with the contents of register No. 1. In sequence 10, register No. 1 was set to 16. So long as the conditional testing is not satisfied, the program will bypass the block immediately following it.

N120,G52,N140 This block transfers the program to sequence 140. This will not occur, however, until 16 holes have been drilled

N130,G52,N60 So long as the number of holes drilled is smaller than 16, the program is redirected to sequence 60 and another hole is drilled.

N140,G29 The 16th hole was drilled, the program reached sequence 140, and the instruction G29 will return the tool to the original starting position

N150,M02 End of program.

This short program allows us to drill any number of holes (only by changing the amount in sequence 10), in a bolt circle of any given diameter (sequence 20). This variable canned cycle, also known as a subroutine, may also be applied to boring, spotfacing, tapping, etc. It would not be used as such, however, with controls, which in addition to variable programming ability, have a bolt circle fixed canned cycle stored in memory.

The above program can be further refined by using two features known as "loopstart" and "loopend" in lieu of the conditional test.

N10,G51,N16 G51 is loopstart, to be "looped" 16 times

N20,F45,G90,G11,R1.875,C0,Z0 Move to hole 1

N30,G91,F4.5,Z-1.2 As in program above

N40,F45,Z1.2 As in program above

N50,G56,C22.5 G56 will cause 22.5° to be added to the contents of register C

N60,G50 Loop ends after 16 holes.

N70,M02 End of program

Because of the investment in conventional CNC equipment, variable canned cycles will not replace the fixed ones overnight, but they establish a clear pattern for the future.

The variable canned cycles, tailored to a particular product or company, have been briefly outlined in this chapter, to allow a closer comparison with the fixed canned cycles supplied as standard equipment or options by the manufacturer.

These subroutines have applications in pocket milling, cam contours and many others. Knowledge of Fortran or Basic programming would be an appreciable asset to the programmer studying variable programming.

Similar programming features are being offered by the Cincinnati Acramatic 900 TC and 900 MC, General Numeric, Fanuc 6MB and 6TB, Sinumerik 8T, and many others.

Applications of variable programming are examined in more detail in Chapter 11.

10

Other
CNC
Features

10.1   SUBROUTINE PROGRAMMING

Subroutines, sometimes called subprograms, are a very powerful time-saving device. They provide the capability of programming certain fixed sequences or frequently repeated patterns and storing them in memory, and provide the opportunity of creating specific canned cycles for a particular application or product. The subroutines can be stored in memory under a specific program number and may be called up by the main program as needed.

The number of subroutines and their size is only limited by the total amount of storage capacity of the control.

Subroutines are in fact independent programs, with their own program numbers, containing all the usual features of a stand-alone program. They are loaded in memory and deleted from memory just like any ordinary program, and, once stored, are viewed by the control in the same manner as a part program.

They have applications in turning and machining centers, as well as other CNC systems.

The subroutine (potentially one of several used in a program), may be called any time as required, and repeated any number of times. Following completion of the required instructions, the command may be returned to the "main" program in the immediately following line of program, or at any other program line specified by the programmer.

e.g., the instruction

N0100 M98 P210

will transfer the command to program number 210, already stored in memory, and containing a given number of tape blocks.

The last block in the subroutine may read M99, following which the command is returned to the line of program immediately following N0100, or it may read, e.g., M99 P0180, at which time the command will be transferred to sequence No. 0180.

In both cases, for the control used in this example, the sequence tape word is designated by the letter P.

The content of the subroutine depends, besides the existing capabilities of the machine-control system, on the experience and the imagination of the programmer.

Some possibilities follow:

1. A complex hole pattern, requiring extensive machining at each hole location, e.g., spot drilling, predrilling, drilling, reaming, grooving (using a special-purpose quill-mounted grooving tool) and chamfering.

    A regular program would change the first tool, set up tool length compensation, perform the machining at all the locations, stop, change the tool and repeat for all the various operations required. Throughout the program, the pattern of hole locations will be observed six times, in this case. Placing the hole pattern in subroutine will mean writing a program containing the locations of all the holes, assigning it a program number, and using it as needed. The programmer will set up a "main" program including only tool changes, spindle and coolant start and stop, as well as the corresponding Z-motions in the appropriate canned cycle. The "main" program will call up and switch to "subroutine" for all the X-Y motions. This will result in a far shorter and overall more efficient program, as well as a better utilization of control storage which affects control response time.

2. A complex contour requiring a number of roughing and finishing milling passes. Again, the entire contour may be located in a "subroutine" program, leading to a much shorter tape.

3. A number of special grooves in a shaft, for which a specific canned (or automatic repeat, or multiple repetitive) cycle may not be available. The groove is programmed in its entirety in subroutine, while the main program will only move the appropriate cutter to the starting point of the next groove.

For the programmer with Basic or Fortran programming knowledge, these subprograms can also be nested, i.e., used repeatedly in a predetermined pattern of instructions.

It is indeed one of the areas in which the selection of a solution largely depends on the imagination of the problem solver.

The following "common sense" notes may be used as a checklist in conjunction with subroutine programming:

- If the main program is in absolute (G90) and the subroutine uses incremental (G91) for its own good reasons, the programmer should reinstate G90 prior to returning the command to the main program.
- If a canned cycle is in effect in the main program and it is not required in subroutine, the latter should start with a canned cycle cancel code. Conversely, should a canned cycle be set up and used in subroutine, it should be canceled prior to returning to main.
- If a cutter diameter compensation or tool nose radius compensation is set up in main, and its transfer to subroutine leads consistently to an error message, the compensation should be set up in subroutine and followed, as prescribed, only by "motion" tape blocks.
- If a subroutine contains a program stop with the corresponding spindle and coolant stops, the latter should be restarted prior to returning to the main program, unless a tool change is involved.
- If the subroutine starts with a return home G28 code, or one is present in the main program return line, the control should be either set to G91 for G28 X0 Y0, or the intermediate point should be selected carefully according to the programming manual instructions.

Essentially, the above notes indicate that errors that would never be made when writing a straight continuous program are quite likely to crop up as the command switches back and forth between different programs. Consequently a more-careful-than-usual dry run is indicated. For example,

Main Program, No. 0001
```
:0001 N0010 . . . . . . . . .
N0020 . . . . . . . . . . . .
N0030 . . . . . . . . . . . .
N0040 P910 M98
```

Command M98 will switch the main program to subroutine, whose own program number is 910, executes its instructions, then return to N0050

```
N0050 . . . . . . . . . . .
N0060 . . . . . . . . . . .
N0070 P910 L2 M98
```
As above, except for the fact that the subroutine instructions will be executed twice, due to L2

```
N0080 . . . . . . . . . .
N0090 . . . . . . . . . . .
          .
          .
          .
          .
N0200 M30
```
End of main program

Subroutine Program, No. 910
```
:910 N0010 . . . . . . . .
```
As the programs under discussion are separate stand-alone programs, there will be no confusion due to similar sequence numbers

```
N0020 . . . . . . . . . . .
N0030 . . . . . . . . . . .
          .
          .
          .
N0100. . . . . . . . . . M99
```
M99 is the command that returns the subroutine to the main program, in the sequence immediately following the one that originated it

If N0100 had read N0100 . . . M99 P0090, the subroutine would have returned to sequence N0090 in the main program. Sequence numbers, usually identified by the tape word "N," are designated as "P" when used in conjunction with subroutine instructions.

10.2   SAFETY (CRASH) ZONE PROGRAMMING

Safety (crash) zone programming, known as stored stroke limit on the General Numerics group of controls, is available for both machining and turning centers.

Programming a safe zone sets up an "invisible" electronic crash barrier, which stops the machine and outputs an alarm when an attempt is made to have a tool penetrate it. This feature is an invaluable capability, since a "crash" that occurs only in the control software is to be preferred by far to an encounter at 400-600 ipm between a tool and a chuck, or a mill and a clamp.

The principle is the same for both milling and turning. In turning, the program will set up an imaginary "box," whose base is a square of side X and whose length is given by Z as follows:

G22 X__Z__I__K__

X, Z, I, and K are all measured from home position (machine zero). X and Z are the coordinates of the corner of the box the closest to machine zero. I and K are the coordinates of the diagonally opposite corner of the box, the farthest from machine zero.

Due to software limitations, the above tape words are to be set in metric only, using trailing zero tape format, even if the rest of the program is in decimal point programming mode.

In milling, the format is similar:

G22 X__Y__Z__I__J__K__

where X, Y, and Z are the coordinates of the corner of the box the closest to machine zero, and I, J, and K, respectively, the farthest coordinates.

Once G22 is set up, any attempt to penetrate the zone will result in a machine stoppage, from which the operator has to back out manually, in the direction exactly opposite to the "offending" motion.

An alternate code is available, G23, which, when combined with a motion statement will allow the tool to penetrate the safety "box."

There is no crash zone cancellation code. The programmer must, in the appropriate tape block, reset the safety zone off the surface of the table, outside of the normal travel limits of the tool. For example,

Note: "Zero" machine and control.
N0010 G80 G40 G49 G20 G00 G90
N0020 G92 X0 Y0 Z4.0
N0030 G22 X10.0 Y-4.0 Z-4.0 I14.0 J-7.0 K-15.0

Sequence N0030 sets up an electronic "box," with its point the closest to machine origin at 10, -4 and -4 inches (X, Y, and Z), and its point the farthest from machine origin at 14, -7, and -15 inches (as given by I, J, and K)

N0040 Z0.5

N0050 G01 Z-0.1 F5.0

N0060 G00 X9.5 Y-4.1   The spindle is now close to the boundary of the electronic box

/N0070 G01 X10.1   If executed, this command will cause an electronic crash

/N0080 G23 G01 X10.1   This is the same instruction as N0070, but the motion will take place without an electronic crash due to G23. The latter allows penetration of the box, but does not cancel it.

N0090 G28 X0 Y0 Z0

N0100 G22 X-1.0 Y1.0 Z1.0 I-2.0 J2.0 K2.0

Sequence N0100 resets the electronic box off the machine table, for all practical purposes canceling it, insofar as this program is concerned.

N0110 M30

11

Introduction to Variable Programming (or "Variable Canned Cycles" or "User Macros" or "Parametric Programming")

To discuss more complex type CNC programming, we must introduce additional codes that allow us to program the machining of contours described by mathematical equations. A continuously changing contour or surface, such as we find on control cams, would require lengthy computer calculations prior to programming without computing capabilities such as described below.

11.1   OPERATORS

=	Equal sign	$(A = B)$
+	Plus sign for addition	$(A + 2)$
−	Minus sign for subtraction	$(A - 3)$
/	Divide operator	$(A / 2.5)$
*	Multiply operator for multiplication	$(C * A)$
,	Comma used for separating words	$(N3 , G59)$
( )	Brackets used for separating groups of operations	

$$\frac{A \cdot B}{\dfrac{3.2 + 4}{2}} = (A * B)/(3.2 + 4)/2$$

[ ]   Square brackets, used to define a variable or a constant in conjunction with the G59 code, such as:

N03, G59, [A] = B
N04, G59, [B] = 3.141593

The value of "B" in N03 may be any variable number, but the value of "B" in N04 is a constant which in future calculations can be used as [B] instead of the 3.141593.

& SI (. . .) Sine (SIN) trigonometric operator (SIN 35.7)
& CO (. . .) Cosine (COS) trigonometric operator (COS 13.978)
& TN (. . .) Tangent (TAN) trigonometric operator (TAN 93.8)
& SR (. . .) Square root ($\sqrt{3.5}$)

NOTE: *The ampersand sign must precede the operator and the values must be enclosed in brackets.*

11.2   SPECIAL INSTRUCTION CODES

G58   Comment code in program, used to provide instructions to the operator in the part program, in the form of:

N035 G01 X3. Y6.4 Z1.2

N040 M01

N045, G58, tool point must be 1.5 inches from part surface.

This message is displayed on the screen of the control

G50 Loop end, used in program as N039, G50

G51 Loop start, used in program as N04, G51, N10

G52 Go, or jump to, used in program as N40, G52, N110

G53 Conditional testing of a calculated quantity against a known quantity. The control will proceed to the next instruction if the value is smaller ($<$) than the known quantity

G54 Conditional testing. The control will proceed to the next instruction if the value is equal ($=$) to the known quantity

G55 Conditional testing. The control will proceed to the next instruction if the value is larger ($>$) than the known quantity

The conditional testing is done to a previously set register as shown below:

N35, G56, D[I]

N39, G55, D90

N41, G52, N110

N43, G52, N05

In sequence N35, the code G56 causes the amount stored at the address "D" to be increased by the amount "I." The value of I had to be established by the programmer prior to this block in the program. G55 in N39 will execute the conditional test. If the value of I is larger than "90" the program will procede to the next block. The next block N41 is an instruction to jump to sequence N110.

(N110 may be the beginning of a new subprogram, the end of the program, or any other CNC instruction.)

If the condition in N39 is equal ($=$) or smaller ($<$), the program will skip the next block and will branch to N43, which is an instruction to return to the beginning, to N05, and repeat the calculations using a different set of values.

This process is repeated until the condition set out by the programmer is found. In this form of programming it is the CNC computer who makes the decision based on the condition set out by the programmer.

G81 Go to subroutine, used in the programs as N35, G81, N1300

Where: N35 is the sequence number of the program, G81 is "go to the address" specified by the following address number; N1300 is the address where the program will continue.

G80   End of subroutine, used in the program as:

      N40, G80

G80   returns the program to the beginning of the subroutine. (Block following the subroutine start.)

G59   Mathematical instruction, used in the program as:

      N15, G59, [3] = A/B

or N20, G59, [A] = 39.7512

The G59 instructs the computer to equate the register inside the square brackets to A/B in sequence N15 and 39.7512 in sequence N20. In the preceding lines, we have selected a few codes of DoALL's Elex Control. We only introduced the ones we need to write our sample programs. The reader who has a Fortran or Basic programming background will find the sample programs presented in the following pages easy to follow. It is not our intention to favor any one system over the other, but to provide sample programs to our readers to use as guidelines in their work. We believe that the CNC systems of the near future will follow this trend of microcomputer numerical control. It will undoubtedly reduce the need for a separate computer for programming. It may also eliminate the use of expensive postprocessors or links.

11.3   DO-LOOP

This is a special CNC programming feature that allows the efficient programming of repetitive machining operations which otherwise would require substantially longer programming and control tape. While the codes for the beginning and the end of the loop may vary from one system to another, the principle, illustrated by the sample programs below, remains the same. The single-loop programming allows us to machine linear arrays, therefore, its use is somewhat similar to the "L" address discussed under the canned cycles. The reader may recall that once the specific canned cycle was programmed, we used the "L" address in the following block as an instruction to repeat the cycle Ln times. The machining in each application was limited to the pattern of the canned cycle. The "Do-Loop" programming allows us to machine practically any shape or form described within the loop. The relevant portion of the program is called the subprogram. Because the programmer may define any required shape within the limits of machinability, the Do-Loop programming is far superior to the canned cycle programming.

The tape formats for our sample program are as follows:

Nm G51 Nn	will start the loop
Where m	is the sequence or tape block number (may be any number from 0000 to 9999) at which the loop starts
G51	is the loop start code
And n	is the number of times the pattern is to be repeated. Its maximum value should be checked in the programming manual
Np G50	will end the loop; G50 is the loop end code; and Np specifies the appropriate sequence number

The CNC system will repeat the operations programmed between the G51 and G50 blocks "n" times.

Example

Write a part program for the machining of the contours in the first line illustrated by Fig. 11-1, then expand the program to repeat the same operations four more times at 3-inch spacing, forming a grid pattern.

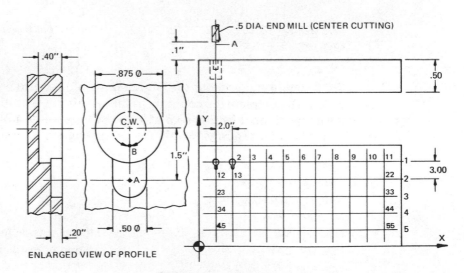

FIG. 11-1. Grid pattern.

ASSUMPTION

Tool is positioned 0.1 inch above the part.

Solution:

N05 G92 X0 Y0 Z0	Zero M/C to control
N10 G91 G00 X2.0 Y13.5	Rapid to point "A"
N15 S400 M03	Spindle on @ 400 rpm
N20 G51 N11	Loop start, 11 repeats
N25 G01 Z-0.3 F5.0 M08	Mill to 0.2 depth at point "A"
N30 Y1.3125	Mill slot to point "B"
N35 Z-0.2	Mill to 0.4 depth
N40 G14 R0.1875 C270 G13 C269.9999	Mill 0.875 blind hole
N45 G00 Z.5	Rapid return to start point
N50 X2.0 Y-1.3125	Rapid to point "2"
N55 G50	Loop end, return to loop start

The CNC reads at sequence number 20 that it must repeat the programmed machining 11 times until it reads loop end in sequence 55. This process is repeated 11 times, then the program continues to sequence 60.

N60 G28 X0 Y0 Z0	Return tool to start point
N65 M02	End of program

This machining, without the Do-Loop, would require 60 additional tape blocks. To complete the contours in the remaining four rows we have to write another loop. Placing the first loop, for the contour milling, inside the second loop for the rows would alter our program as shown below:

N05 G92 X0 Y0 Z0	
N10 G91 G00 X2.0 Y13.5	
N15 S400 M03	
N20 G51 N5	First or outside loop start
N25 G51 N11	Second or inside loop start
N30 G01 Z-0.3 F5.0 M08	
N35 Y1.3125	
N40 Z-0.2	

```
N45 G14 R0.1875 C270 G13 C269.9999
N50 G00 Z.5
N55 X2.0 Y-1.3125
N60 G50                         End of second or inside loop
```

The program will return to N25, the start of the inside loop, as many as, in our example, 11 times, as the program calls for, causing locations 1 to 11 to be machined. Then it continues to sequence 65.

```
N65 G00 X-20.0 Y-4.3125      Rapid to point 12
N70 G50                      End of first or outside loop
```

The program returns to the outside loop start sequence, N20, which will reinitiate the inside loop. The second time, in the inside loop, locations 12 to 22 will be machined. This process is repeated 5 times to complete the contour milling at all 55 locations. Then it continues to sequence N75.

```
N75 G00 G28 X0 Y0 Z0         End of program
N80 M02
```

Conclusion

By using conventional programming methods, this program would require 390 tape blocks to machine the sample part.

Tape or program proving (try out) takes one sixty-fifth of the conventional CNC program as the program itself only represent 6 blocks of tape. The balance consists of repeats performed by the CNC computer, therefore, the possibility of programming error is eliminated.

EXERCISE

Write a part program to machine the multiple circular pattern illustrated by Fig. 11-2. The elongated hole contour is to be repeated in 36 equally spaced locations on a 10.00-inch base circle. The part has three hole circles 12.50 apart in the X-direction of the machine axis.

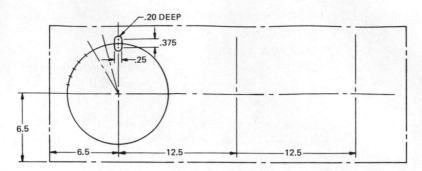

FIG. 11-2. Multiple circular pattern.

11.4 MATHEMATICAL INSTRUCTIONS

The part programming of a three-dimensional contour milling, such as an internal hemisphere, would require lengthy calculations prior to the actual program writing. Using the mathematical instructions, we can combine the calculations with the tool motion commands in one program.

Example

Write a part program for the milling of the internal hemisphere illustrated by Fig. 11-3. The program will have to be used for a family of parts with the minimum possible change to the program.

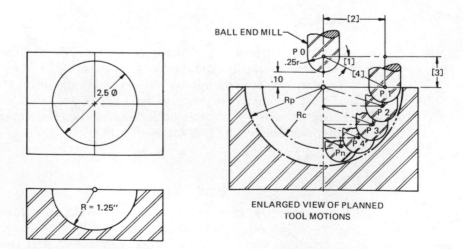

FIG. 11-3. Internal hemisphere.

SOLUTION

1. Sketch an enlarged view (as shown in Fig. 11-3) of the part, establish the pattern of tool motions, and label the variables.

2. Show tool start point on center line of part, 0.1 inch above part surface.

3. Define machining pattern; move tool to point P1 and cut a 360° arc at the level of this point.

4. Repeat preceding step at points P2, P3, etc., until the internal hemisphere is completed.

5. Establish the stepping angle [1] to provide the required surface finish.
   [1] = 3°. Should the surface finish be unsatisfactory you may reduce this value to 2° or 1° or any smaller fraction of the above angle.

6. Identify the rest of the dimensions as shown below:

[9] = 0.25	Radius of end mill
[8] = 1.25	Radius of part
[4] = [8] − [9]	Radius of end mill center
[7] = 0	Zero will be the amount of the starting angle
[2] = & co([7] * [4])	X-tool motion
[3] = & si([7] * [4])	Z-tool motion

NOTE: *We would normally not label the part and the tool radii as variables. However, if the program will be used for a family of parts with different radii, the changes could be limited to [9] and [8].*

7. Write the part program:

N05 G58, Main program for milling of hemisphere
N10 G58, Part radius = 1.25, ball end mill radius = 0.25
N15 G59, [9] = 0.25
N20 G59, [8] = 1.25
N25 G59, [4] = [8] − [9]
N30 G59, [1] = 3.
N35 G59, [7] = 0
N40 G92, X0 Y0 Z-0.35
N45 M03 M10 G04 N500   Spindle and coolant on, 5-sec dwell
N50 G59, [2] = & co([7] * [4])   Calculate the X-component of the first motion

N55 G59, [3] = & si([7] ∗ [4])   Calculate the Z-component of the first motion

N60 F8. G90 G01 X [2] Y0. Z-[3]   Move the tool to the calculated X- and Z-location

N65 G91 G14 R [2] C0 G13 C360   Where G14 identifies the present tool point as the starting point of the circle in the polar coordinate system.
R[2] The calculated radius of the circle at point P1.
C0 This is the starting angle on the circle.
G13 Directs the tool motion in the CCW direction.
C360 This is the finish point of the tool motion. At this point, our tool has performed a 360° circular motion above the part.

N70 G59, [7] = [7] + [1]   Increment the original angle (set to zero) by the stepping angle (3°) for the second round of calculations and circular machining.

N75 G56 D [1]   Increment the "D" register by the value of [1]. This register will serve in our program for the conditional test. To complete the milling of the hemisphere, the angle will have its value progressively increased to 90°.

N80 G55 D90   Conditional test. If the value of D [1] is larger than D90, the program will continue to the next tape block, meaning that the milling is completed. Otherwise it will skip the following tape or program block.

N85 G52 N95

N90 G52 N50

Since the value of D [1] *is not larger than the test value, the program will prompt to sequence N90. This is an instruction to return to N50 and continue the calculations with the new angular value indexed in N70. The process will be repeated until the condition set out by the programmer is reached. This condition is for D* [1] *to be larger than 90°. When this condition is reached, the program will go to the following tape block, N85. This block will now instruct the computer to go (jump) to N95.*

N95 G29   The tool will return to the start point.
N100 M05   Spindle stop.
N110 M02   End of program.

Conclusion

The part programming of an internal hemisphere, often used in the die industry, would have required computer assistance or lengthy calculations by the programmer in conventional CNC programming. For the cutting motions only, the program would require thirty linear motions from P1 to P2, etc., to Pn, and thirty circular interpolations of 360°, for a total of 60 blocks. The conventional CNC program would not allow for adjustment should the surface finish prove to be unsatisfactory.

These two points alone will clearly indicate the superiority of the customer-written subroutine programming. In lieu of the above 60 blocks, we have programmed the machining of the part in only 21 blocks. The flexibility is built into the program using the statement [1] = 3. Should the surface finish, after the trial cut, prove to be unsatisfactory, all the programmer has to do is to change the 3° value to a smaller angle. The second objective of our example was to write the program for a family of parts. This was implemented by using variables in the tool motion blocks N60 and N65.

Should, for instance, the part radius of the second part in the family be 1.5 inches and the tool radius be 0.3125 inch, the only alteration the programmer has to do is change [9] = 0.25 to [9] = 0.3125 and [8] = 1.25 to [8] = 1.5 in program sequences N15 and N20.

In most plants, the programmer can provide this information to the supervisor or operator who can edit the program to suit any part or tool dimension in the family. No other change to the program would be needed since the tool motions X [2] and Z- [3] in tape blocks N60 and N65 are calculated in terms of [1], [9], and [8] by the minicomputer CNC.

EXERCISE

Write a part program for the milling of the sample part illustrated in Fig. 11-4.

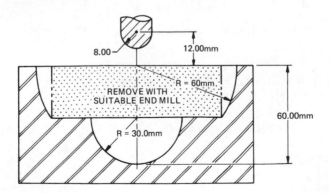

FIG. 11-4. Compound contour.

11.5   SUBPROGRAM WITH DO-LOOPS

The programming examples discussed so far were relatively easy. The programmer was able to visualize the mathematical as well as the machining cycles without lengthy notes and could follow them both separately and together. This task may become far too complicated, perhaps impossible, if the contouring job requires complex mathematical do-loops and subroutines in addition to the tool motions. Such is the case in our next example. The sample part illustrated in Fig. 11-5 is a mechanical cam. Its function is to guide a forming die on a special-purpose machine. The circular path of the cam follower will be identical to the tool path of our program.

The cam has two constant radius sections, from point "B" to point "C" and from point "D" to point "A." The section from point "A" to point "B" represents the acceleration and from point "C" to point "D" the deceleration. The change in radius from point "A" (R3.25) to point "B" (R4.25) in acceleration is described by the following mathematical equation which will also be used in the deceleration curve:

$$R = R_0 + \frac{S}{2 + \pi} \left(\frac{2I}{T} - \frac{1}{2\pi} \cdot \sin\left(\frac{720 \cdot I}{T}\right) \right) \quad \ldots \ldots \text{11.5a}$$

This formula is valid between the angles of $0 \geqslant I \geqslant T/8$, or $0 \geqslant I \geqslant 8.125°$, where T is the total angle of change (65°)

S is the change in radius from point "A" to point "B" (1.0 inch)
I is the change of angle (incremental)
$\therefore$ The tool path will change as I is incremented from $I = 0$ to $I = T/8$
$(65/8 = 8.125°)$

$$R = R_0 + \frac{S}{2 + \pi} \left(0.25 - \frac{1}{2\pi} + \frac{2}{T}\left(I - \frac{T}{8}\right) - \frac{4\pi}{T^2}\left(I - \frac{T}{8}\right)^2 \right) \quad \ldots \text{11.5b}$$

This formula is valid between the angles of $T/8 \geqslant I \geqslant 3T/8$, or $8.125° \geqslant I \geqslant 24.375°$.

$$R = R_0 + \frac{S}{2 + \pi}\left(-\frac{\pi}{2} + 2\left(\frac{1 + \pi}{T}\right) \cdot I - \frac{1}{2\pi}\sin\left(\frac{720}{T} \cdot I - 180\right) \right) \ldots$$
$$\text{11.5c}$$

This formula applies for $3T/8 \geqslant I \geqslant 5T/8$, or $24.375° \geqslant I \geqslant 40.625°$.

$$R = R_0 + \frac{S}{2 + \pi}\left(1.75 + \frac{1}{2\pi} + \pi - \frac{2}{T}\left(\frac{7T}{8} - I\right) - \frac{4\pi}{T^2}\left(\frac{7T}{8} - I\right)^2 \right) \ldots$$
$$\text{11.5d}$$

The domain of this formula is $5T/8 \geqslant I \geqslant 7T/8$ or $40.625° \geqslant I \geqslant 56.875°$.

$$R = R_0 + \frac{S}{2 + \pi}\left(\pi + \frac{2I}{T} - \frac{1}{2\pi}\sin\left(\frac{720I}{T}\right) \right) \quad \ldots \text{11.5e}$$

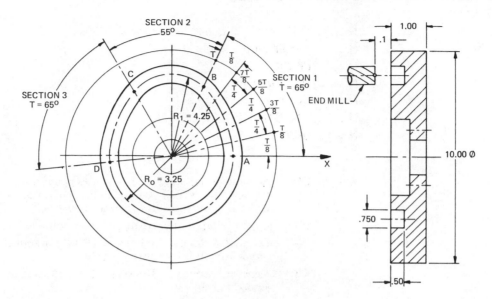

FIG. 11-5. Mechanical cam.

This is the final section of the curve corresponding to $7T/8 \geqslant I \geqslant T$, or $56.875° \geqslant I \geqslant 65°$.

At first glance, the problem looks complex enough even for an experienced programmer. However, if we carefully analyze the task and break it down into simple elements, it becomes relatively easy.

Prior to proceeding with our program, we should make two additional notes. The Elex control on the DoALL recognizes the digits 0, 1, 2 . . . 9 and any alphabetic character not used in the main program as usable variables. In addition, it has no character for π which will have to be handled differently.

SOLUTION

1. Analyze the problem and its possible solutions.

The tool path must be calculated by equation 11.5a from zero to 8.125°. If we divide the 8.125° into small enough elements, e.g., 8.125/26 = 0.3125°, and we substitute this increment into our equation for I, we can then write a motion statement using the angle and the radius as polar coordinates in the same tape block. To move the tool after the first motion, we add 0.3125°, we repeat the calculations, and we move the tool to the new coordinates. The microcomputer CNC will repeat the process 26 times to complete the first section.

That program for the above, would look like the following:

N . . .   G51, N26   Loop start, repeat 26 times.

N . . .   increment the angle

N . . .⎤ do the

N . . .⎬ calculations

N . . .⎦ as per equation 11.5a

N . . .   G90 G11 R [8] C [5] Z-0.6 F5

N . . .   G50

Where   G90   Absolute programming. The values generated
                mathematically will be used in the motion blocks
                while keeping track of the starting motion.

        G11   Linear interpolation using polar coordinates.

        R[8]   Calculated radius. We have tentatively assigned the
                number eight [8] for this variable.

        C[5]   Angular increment. The number five [5] was as-
                signed to this variable.

        Z-0.6 This is the depth of the cam track plus 0.1-inch tool
                clearance. We must have this information in the
                tape block because of the absolute mode of program-
                ming.

        G50   Will keep sending the program back to the loop
                start until all 26 loops are completed.

2.   To program the second segment from $8.125°(T/8)$ to $24.375°(3T/8)$
     we must change the starting angle from zero (0) to $8.125°$ then pro-
     ceed with the calculations using equation 11.5b. If we assign a
     variable to this angle in the form of [9], then we can use the same
     notation in our motion statement. The whole process will have to
     be repeated three more times for the remaining zones. It is there-
     fore advisable to place the do-loop into a subroutine such as:

N . . .   G59, [3] = 8.125

N . . .   G81 N . . .   Subroutine starts

N . . .

N . . .

N . . .   G51 N52   Loop starts

N . . .   Increment the angle

N . . .⎤ do the

N . . .⎬ calculations

N . . .⎦ as per equation 11.5b

N . . .   G90 G11 R [8] C [5] Z-0.6

N . . .   G50   Loop ends

N . . .   G80   Subroutine ends

The "N" address following the G81 within the same block will transfer the program to the second do-loop start block. This block will instruct the computer to repeat the calculations and tool motions 52 times (i.e., 52 × 0.3125 = 16.250° which is the angle between 24.375° and 8.125°). Once the loop is completed, the G80 will return the program to the block following the G81 code.

3. Repeat the entire procedure for equations 11.5c, 11.5d, and 11.5e.

4. Use the same "subroutine over do-loop" programming technique to complete the deceleration section from point "C" to point "D." The same equations describe the tool path in the deceleration cycle. For most cams, the acceleration and deceleration angles are different, and the programmer must then establish angular values for each segment separately, for both acceleration and deceleration section.

5. Assign variables: Any value which may be used repeatedly should be handled as a variable. This will substantially reduce the length as well as the complexity of the program.

Starting angle: G59, [3] = 0   in segment one
We will assign different values for this address in each segment as the program will require:

	G59, [3] = 8.125	in segment two
	G59, [3] = 24.375	in segment three
	G59, [3] = 40.625	in segment four
	G59, [3] = 56.875	in segment five
Starting radius:	G59, [4] = 3.25	in section one
and	G59, [4] = 4.25	in section three

Angular increment: G59, [6] = 0.3125
Total angle of acceleration and/or deceleration:
                       G59, [1] = 65

Total angle for motion:

	G59, [9] = 0	in section one
and	G59, [9] = 65	in section three

Rise and fall of radial displacements:

	G59, [S] = 1.	for acceleration
and	G59, [S] = −1.	for deceleration
As π = 3.141593	G59, [W] = 3.141593;	variable W will represent π.

6. As the given equations are fairly long, we will endeavor to simplify them by assigning variables to terms that are common in more than one equation. This process will reduce the corrections to the program. Long, complicated equations often lead to programming errors, which are difficult to notice.

$$\frac{S}{2 + \pi} \ldots\ldots\ldots\ldots \text{G59, [P]} = \text{[S]} /(2 + \text{[W]})$$

$$\frac{1}{2\pi} \ldots\ldots\ldots\ldots \text{G59, [Q]} = 1/(2 * \text{[W]})$$

$720 \cdot \text{I/T} \ldots\ldots\ldots \text{G59, [U]} = (720 * \text{[3]})/\text{[1]}$

$2 \cdot \text{I/T} \ldots\ldots\ldots \text{G59, [V]} = 2 * \text{[3]}/\text{[1]}$

$4 \cdot \pi/\text{T}^2 \ldots\ldots\ldots \text{G59, [A]} = (4 * \text{[W]})/(\text{[1]} * \text{[1]})$

$\text{I} - \text{T/8} \ldots\ldots\ldots \text{G59, [B]} = \text{[3]} - (\text{[1]}/8)$

$2(1 + \pi/\text{T}) \cdot \text{I} \ldots\ldots \text{G59, [L]} = ((1 + \text{[W]})/\text{[1]}) * 2 * \text{[3]}$

$(720 \cdot \text{I})/\text{T} - 180 \ldots\ldots \text{G59, [M]} = ((720 * \text{[5]})/\text{[1]}) - 180$

$7\text{T/8} - \text{I} \ldots\ldots\ldots \text{G59, [Q]} = (7 * \text{[1]})/8 - \text{[3]}$

$(720 \cdot \text{I})/\text{T} \ldots\ldots\ldots \text{G59, [H]} = (720 * \text{[3]})/\text{[1]}$

$(7 \cdot \text{I})/\text{T} - \text{I} \ldots\ldots\ldots \text{G59, [Z]} = (7 * \text{[1]})/8 - \text{[3]}$

7.   Writing the part program:
     The reader should carefully study each statement and should not proceed into the program without understanding every element of a statement or instruction. *In this control, commas or spaces are valid word separators.*

12345	Program number
N5, G58, sample program	Program title, displayed on screen.
N10, G92, X0, Y0, Z0	Zero machine and control
N15, F25., G90, G11, R3.25, C0, Z0	Move tool to point "A"
N20, G01, Z-.6, F4.5	Cut to 0.5 depth
N25, G59, [1] = 65	Total acceleration angle.
N30, G59, [3] = 0	Starting angle
N35, G59, [4] = 3.25	Starting radius
N40, G59, [W] = 3.141593	Value of π
N45, G59, [9] = 0	Angle of tool location
N50, G59, [6] = .3125	Angular increment
N55, G59, [S] = 1.	Rise of cam

All variables are now set for the programming of the first element using equation 11.5a. At this point we can transfer the program to sequence N1000 which has been selected for the start of the subprogram.

N60, G81, N1000	First subprogram starts at N1000
N65, G59, [3] = 8.125	Starting angle for second element
N70, G81, N1100	Second subprogram starts at N1100

N75, G59, [3] = 24.375	Starting angle for third element
N80, G81, N1200	Third subprogram starts at N1200
N90, G59, [3] = 40.625	Starting angle for fourth element
N95, G81, N1300	Fourth subprogram starts at N1300
N100, G59, [3] = 56.875	Starting angle for final element
N105, G81, N1400	Final subprogram starts at N1400
N110, G52, N1450	Jump to sequence N1450
N1000, G51, N26	Do-loop starts, 26 repeats
N1005, G59, [3] = [3] + [6]	Starting angle 0 + 0.3125°
N1010, G59, [P] = [S] /(2 + [W])	$S/(2 + \pi)$
N1015, G59, [Q] = 1/(2 * [W])	$1/2\pi$
N1020, G59, [U] = (720 * [3])/[1]	$720 \cdot I/T$
N1025, G59, [V] = 2 * [3]/[1]	$2 \cdot I/T$
N1030, G59, [8] = [4] + [P] * ([V] − [Q] * (&si([U]))) Cam radius is 8	
N1035, G59, [5] = [3] + [6] + [9]	Angle of motion
N1040, G90, G11, R [8] , C [5] , Z-0.6	Motion in polar coordinates
N1045, G50	Loop ends
N1050, G80	Subroutine ends

The program now returns to the block following the subroutine start N65, assigns 8.125 to the new starting angle, and continues with N70. The G81 in N70 is the start of the second subroutine and the program proceeds with N1100.

N1100, G51, N52	Do-loop starts, 52 repeats
N1105, G59, [3] = [3] + [6]	8.125 + 0.3125
N1110, G59, [A] = (4 * [W])/([1] * [1])	$4\pi/T^2$
N1115, G59, [B] = [3] − ([1]/8)	$I - T/8$
N1120, G59, [8] = [4] + [P] *(0.25 − [Q] + ((2 * [B])/[1]) + ([A] * [B] * [B]))	
N1125, G59, [5] = [3] + [6] + [9]	
N1130, G90, G11, R [8] , C [5] , Z-0.6	
N1135, G50	Do-loop ends
N1140, G80	Subroutine ends

The program now returns to the block following the subroutine start N75, assigns 24.375° to the new starting angle, and continues with N80. The G81 in N80 is the start of the third subroutine which sends the control to N1200. This process is repeated after each subroutine until the acceleration cycle is completed.

N1200, G51, N52

N1205, G59, [3] = [3] + [6]

N1210, G59, [L] = ((1 + [W])/[1]) * (2 * [3])

N1215, G59, [0] = ((720 * [3])/[1]) − 180

N1220, G59, [8] = [4] + [P] * ((−[W]/2) + [L] − ([Q] * (& si([0]))))

N1225, G59, [5] = [3] + [6] + [9]

N1230, G90, G11, R [8] , C [5] , Z-0.6

N1235, G50

N1240, G80

N1300, G51, N52

N1305, G59, [3] = [3] + [6]

N1310, G59, [2] = (7 * [1])/8 − [3]

N1315, G59, [8] = [4] + [P] * (1.75 + 1/(2 * [W]) + [W] − (2 * [2])/[1] − −((4 * [W])/([[1] * [1]) * [2] * [2]))

N1320, G59, [5] = [3] + [6] + [9]

N1325, G90, G11, R [8] , C [5] , Z-0.6

N1330, G50

N1335, G80

N1400, G51, N26

N1405, G59, [3] = [3] + [6]

N1410, G59, [H] = (720 * [3])/[1]

N1415, G59, [8] = [4] + [P] * ([W] + (2 * [3])/[1]) − [Q] * & si ([H]))

N1420, G59, [5] = [3] + [6] + [9]

N1425, G50

N1430, G80

At this point, the acceleration curve has been completed to point "B." The program will transfer back to the block following the last subroutine start. This block (N110) contains a G52 (jump to) which unconditionally transfers the control to N1450.

N1450, G14, R4.25, C65 , G13, C120

This block will move our cutter to point "C" on a 4.25 constant radius. Prior to starting the decelertion curve we have to reassign appropriate values to the following variables.

N1455, G59, [3] = 0

N1460, G59, [4] = 4.25

N1465, G59, [1] = 65

N1470, G59, [9] = 65
N1475, G59, [S] = −1.
N1480, G81, N1000

The [S] = −1. will change the sign of all five equations to the right of the R_0. These calculated values will be subtracted from the new radius [4] = 4.25 for each calculated radius in the program. The G81 will start the first subroutine in block N1480, and the deceleration curve will be calculated and machined to point "D."

N1485, G59, [3] = 8.125
N1490, G81, N1100
N1495, G59, [3] = 24.375
N1500, G81, N1200
N1505, G59, [3] = 40.625
N1510, G81, N1300
N1515, G59, [3] = 56.875
N1520, G81, N1400

The deceleration curve has now been completed to point "D." The remaining section to point "A" has a constant radius of 3.25. Therefore, it does not require any specific calculations.

N1525, G14, R3.25 , C [5] , G13, G0
N1530, G29
N1535, M02

Conclusion

The preceding example has established a pattern of part programming for mechanical cams with a cycle pattern of acceleration-constant velocity-deceleration-constant velocity. Should the programming require multiple cycle acceleration and deceleration, as well as the addition of tool offsets, the program would have to be rewritten. Adding the tool offset will provide us with flexibility of using any size of end mill, with a diameter smaller than the width of the cam track. The following program of a three-cycle cam will illustrate the technique of programming the multiple cycle cams.

EXAMPLE

Write a part program for the machining of the sample mechanical cam illustrated in Fig. 11-6. The program is to be written with cutter diameter compensation (tool offset) to allow for any size of end mill with a diameter smaller than 0.750.

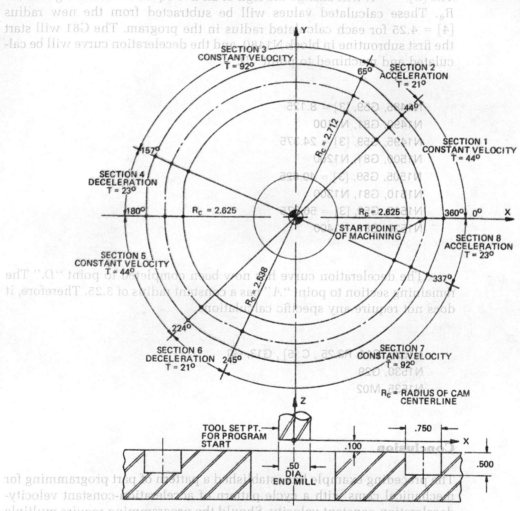

FIG. 11-6. Multicycle mechanical cam.

Design Data

1. The centerline of cam track has a constant radius of 2.625 inches between the angles of 0° and 44°.

2. The centerline radius of the cam track varies from 44° to 65°. The rise is 2.712 − 2.625 = 0.87 inch = S. This change in the curve must occur in accordance with the equations introduced in the previous pages. The rate and limits of the changes are described below.

Between the angles of $0 \geqslant I \geqslant T/8$ or $0 \geqslant I \geqslant 2.625°$, the tool path is defined by the following equation of the radius R:

$$R = R_0 + \frac{S}{2 + \pi}\left(\frac{2I}{T} - \frac{1}{2\pi} \cdot \sin\left(\frac{720 \cdot I}{T}\right)\right), \text{ where } T = 21°$$

Notations are identical to our previous example. The reader should notice that this equation is identical to 11.5a. However, the rate of change (S) and the angle over the changing section (T) are different. Between the angles of $T/8 \geqslant I \geqslant 3 T/8$ or $2.625° \geqslant I \geqslant 7.875°$, the tool path is defined by equation 11.5b.

Between the angles of $3 T/8 \geqslant I \geqslant 5 T/8$ or $7.875° \geqslant I \geqslant 13.125°$, the tool path is described by equation 11.5c.

Between the angles of $5 T/8 \geqslant I \geqslant 7 T/8$ or $13.125° \geqslant I \geqslant 18.375°$, the tool path is given by equation 11.5d.

Between the angles of $7 T/8 \geqslant I \geqslant T$ or $18.375° \geqslant I \geqslant 21°$, the tool path equation is given by the formula 11.5e.

3. The center line of the cam track has a second constant radius of 2.712 inches over a 92° arc from 65° to 157°.

4. The center line of the cam track will vary in the opposite direction from 157° to 180°. The variation is 2.625 − 2.712 = −0.087 inch = S. However, the change in the curve will be guided by different parameters: $T = 23°$ and $S = -0.087$. Using the same equations, 11.5a to 11.5e, the limits for each equation are shown below:

$0° \geqslant I \geqslant T/8$
$0° \geqslant I \geqslant 2.875°$ for equation 11.5a,

$T/8 \geqslant I \geqslant 3T/8$
$2.875° \geqslant I \geqslant 8.625°$ for equation 11.5b,

$3T/8 \geqslant I \geqslant 5T/8$
$8.625° \geqslant I \geqslant 14.375°$ for equation 11.5c,

$5T/8 \geqslant I \geqslant 7T/8$
$14.375° \geqslant I \geqslant 20.125°$ for equation 11.5d and

$7T/8 \geqslant I \geqslant T$
$20.125° \geqslant I \geqslant 23°$ for equation 11.5e

The remaining four sections of the cam are identical to the first four, therefore, the rate of change in sections 6 and 8 are described above under 1 and 3, using opposing signs for the variable "S."

Solution

Prior to proceeding with our part program we must outline some of the changes we had to make in the programming.

1. Since one of the programming objectives was to use cutter diameter compensation (tool offsets), we have to program a tool code (T1). The tool code will allow the operator to insert the appropriate offset value (0.125 inch) for the 0.500-inch diameter tool. This offset value is the cutter radius differential.

2. The five equations describe the positional change in the cam track center line. Adding or subtracting the value of the offset to the starting radius of 2.625 inches will not produce the required cam.

3. As a result, we had to change the tool motion blocks from polar to cartesian coordinates. We have accomplished this by calculating the appropriate radii [8] as a function of the angular increment [5], just as in the previous example. Using these values, we have calculated the actual X- and Y-locations for each angular step, then we programmed the tool motion in terms of those X-Y values. These programming changes will produce the required cam track. Replacing the 0.5-inch-diameter end mill with a spring-loaded pencil in the spindle, we produced a tool center line plot on the machine (see Fig. 11-7), prior to machining the cam. The actual cam was used for the rotation of a link on a chain-link fabricating machine.

4. The Part Program:

```
7777                              Program number
N5, G58, Chain link cam 7
N10, G92, X0, Y0, Z0
N11, G58, Set tool, as shown in Fig. 11-6
N15, F25
N20, T1
N25, G41
N30, G90, G01, X2.625, Y-1., M03
N35, Y0.
```

The Y-motion in N30 was required to activate the cutter diameter compensation (tool offset) G41 from sequence N25.

```
    N40, G90, G11, R2.625, C0, Z0      This block changes the control
                                       mode to polar coordinates
                                       without any tool motion.
```

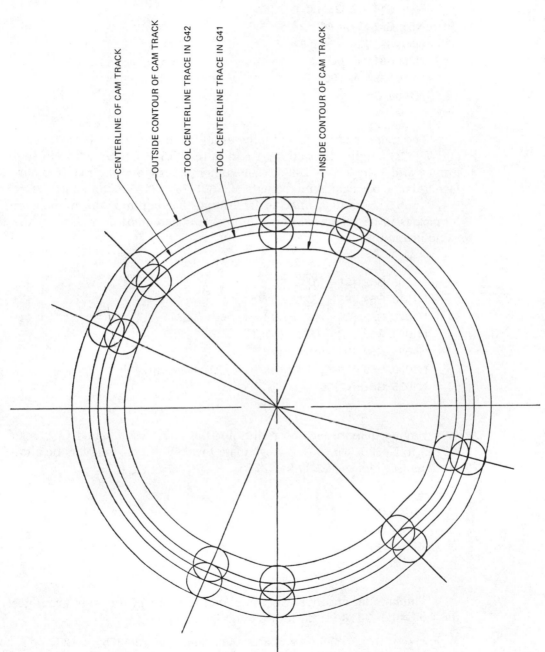

CENTERLINE OF CAM TRACK

OUTSIDE CONTOUR OF CAM TRACK

TOOL CENTERLINE TRACE IN G42

TOOL CENTERLINE TRACE IN G41

INSIDE CONTOUR OF CAM TRACK

FIG. 11-7. Verification of cam contour using CNC mill as a plotter.

213

```
N45, Z-0.6, F4.5
N50, G14, R2.625, C0, G13, C44
N55, G59, [9] = 44
N60, G59, [S] = -0.087
N65, G81, N2100                    Start subprogram in N2100
N125, G52, N1500                   Jump to block 1500
N1000, G51, N6                     Loop starts
```

The number of loops had to be established in view of both $T = 21°$ and $T = 23°$ angles. In sections 1 and 6, the $T/8 = 2.625°$; while in sections 4 and 8 the $T/8 = 2.875°$. The six loops provided the required surface finish to our cam at increments of $2.625°/6 = 0.4375°$ and $2.875°/6 = 0.47916667°$. Should a finer surface finish be required, the number of loops has to be increased and the angle of increment decreased in the same proportion.

```
N1005, G59, [3] = [3] + [6]
N1010, G59, [P] = [S]/(2 + [W])
N1015, G59, [Q] = 1/(2 * [W])
N1020, G59, [U] = (720 * [3])/[1]
N1025, G59, [V] = 2 * [3]/[1]
N1030, G59, [8] = [4] + [P] * ([U] - [Q] * (& si ([U])))
N1035, G81, N2200
```

Having determined the radius for the angle established in blocks N2100 to N2120, we can now calculate the X-Y dimensions for the slide motions (see blocks N2200 to N2215).

```
N1040, G50
N1045, G80
N1100, G51, N12
```

The number of loops was calculated in terms of the angular ranges for $T/8 \geq I \geq 3T/8$ for all four changing sections.

For $T = 21°$: $7.875° - 2.625° = 5.25°$ and $5.25°/0.4375° = 12.$
For $T = 23°$: $8.625° - 2.875° = 5.75°$ and $5.75°/0.47916667° = 12.$

N1105, G59, [3] = [3] + [6]

N1110, G59, [U] = (4 * [W])/([1] * [1])

N1115, G59, [V] = [3] − ([1]/8)

N1120, G59, [8] = [4] + [P] * (0.25 − [Q] + ((2 * [V])/[1]) +
+([U] * [V] * [V]) .

N1125, G81, N2200

N1130, G50

N1135, G80

N1200, G51, N12

N1205, G59, [3] = [3] + [6]

N1210, G59, [V] = ((1 + [W])/[1]) * (2 * [3])

N1215, G59, [O] = ((720 * [3])/[1])) − 180

N1220, G59, [8] = [4] + [P] * ((−[W]/2) + [V] − ([Q] * (& si (0))))

N1225, G81, N2200

N1230, G50

N1235, G80

N1300, G51, N12

N1305, G59, [3] = [3] + [6]

N1310, G59, [O] = (7 * [1])/8 − [3]

N1315, G59, [7] = ((4 * [W])/([1] * [1])) * ([O] * [O])

N1320, G59, [8] = [4] + [P] * (1.75 + 1/(2 + [W] + [W] −
−((2 * [O])/[1] − ([7]))

N1325, G81, N2200

N1330, G50

N1335, G80

N1400, G51, N6

N1405, G59, [3] = [3] + [6]

N1410, G59, [O] = (720 * [3])/[1]

N1415, G59, [8] = [4] + [P] * ([W] + ((2 * [3])/[1]) − ([Q] * & si ([O])))

N1420, G81, N2200

N1425, G50

N1430, G80

N1500, G14, R [8] , C [5] , G13, C157

N1510, G59, [9] = 157

N1520, G59, [S] = 0.087

N1530, G59, [4] = 2.538

N1540, G81, N2300

N1900, G14, R [8] , C [5] , G13, C224

N1910, G59, [9] = 224

```
N1920, G59, [S] = 0.087
N1930, G81, N2100

N2000, G14, R [8] , C [5] , G13, C337
N2010, G59, [9] = 337
N2020, G59, [S] = − 0.087
N2030, G59, [4] = 2.712
N2040, G81, N2300
N2050, G52, N2400

N2100, G59, [W] = 3.141593
N2105, G59, [3] = 0
N2110, G59, [4] = 2.625
N2115, G59, [1] = 21
N2120, G59, [6] = 0.4375
N2125, G81, N1000
N2130, G59, [3] = 2.625
N2135, G81, N1100
N2140, G59, [3] = 7.875
N2145, G81, N1200
N2150, G59, [3] = 13.125
N2155, G81, N1300
N2160, G59, [3] = 8.375
N2165, G81, N1400
N2170, G80

N2200, G59, [5] = [3] + [9]
N2205, G59, [X] = [8] ∗ (& co ([5]))
N2210, G59, [Y] = [8] ∗ (& si ([5]))
N2215, G90, G01, X [X] , Y [Y]
N2220, G80

N2300, G59, [1] = 23

N2305, G59, [3] = 0
N2310, G59, [6] = 0.47916667
N2315, G81, N1000
N2320, G59, [3] = 2.875
N2325, G81, N1100
N2330, G59, [3] = 8.625
N2335, G81, N1200
N2340, G59, [3] = 14.375
N2345, G81, N1300
N2350, G59, [3] = 20.125
N2355, G81, N1400
```

N2360, G80
N2400, G00, Z0
N2410, G40
N2415, Y-0.25
N2500, G29
N2600, M02

This program produced the inside profile of the cam on the first pass. For the second pass, the cutter diameter compensation left (G41) was switched to a cutter diameter compensation right (G42) in block N20 (using the same offset value 0.125), and the tool produced the outside profile of the cam track.

Conclusion

In these programs we have illustrated the powerful programming features of the minicomputer numerical control. This program could not have been written without the use of a computer-assisted language built into a conventional CNC. The customer-written subprograms (user macros) can be custom-tailored for utmost efficiency. The program has substituted different angular values into complex mathematical equations, calculated the required radii, and moved the tool almost simultaneously. The program provides the added flexibility to produce the cam track with any undersized cutter by changing the value of the offset (cutter radius differential) in the offset register. It is becoming quite apparent that the only limitation this system may have is in the programmer's ability. The one drawback is that the programmer must learn a few additional codes in addition to the conventional G-codes discussed in the previous chapters. To further facilitate the programmer's task, special computer-assisted languages have been written. Symbolic English-like instructions in a computer-assisted part programming language can do all the coding error free. Additional linkage to a translator called "postprocessor" will tailor the coding to the specific machine on which the part will be made. This concept of programming will be discussed in detail in chapter 12.

12

Introduction to Computer-Assisted Part Programming

In the previous chapters we have explored the many intricacies of manual part programming. The term "manual" applies to all situations in which the program is written in machine codes, even though microcomputers may have been used to prepare and process the tape. The term "computer-assisted" represents the use of a specific language, to be discussed later in this chapter.

In spite of the efforts of the manufacturers and the various professional organizations, the CNC industry has so far been unable to come up with a compatible coding standard. This was primarily because no national or international body has the power to regulate or enforce the uniform use of codes for manual programming. As a result, the same "G" code may represent three different functions on three different systems. Programs written for one control type cannot be used for another, even if the machine is the same kind and size, because of incompatible codes. The authors have encountered incompatibility even when the machine and control were the same, because additional features had been built into one of the systems, somewhat more recent than the other. For this reason, programs often had to be rewritten—at added cost—if jobs had to be shifted from one machine to another. Alternatively, the company loses flexibility in the area of shop scheduling.

In addition to coding, manual part programming requires accurate and consistent calculations, occasionally far more extensive than the coding itself.

In chapter 11, a recent concept of manual programming was introduced and discussed, involving additional codes and programming rules, and providing extensive flexibility and efficiency. This user-written subroutine (macro) programming requires additional knowledge in trigonometry and analytical geometry. It also, unfortunately, carries the same burden as its predecessors: the programming formats for the mathematical elements are incompatible.

Manufacturers introduce new programming features yearly, independent of each other, and the result is that the same problem is programmed differently on each control. A programmer who programs manually several totally different machine-control systems is prone to make costly mistakes in terms of scrapped parts, broken tools, and machine damage. These situations delay production schedules and add to the cost of the manufactured product. Before getting involved in our discussion of computer-assisted systems, we should emphasize that manual programming is performed effectively and efficiently for two-axis contouring applications. Relatively simple parts, with linear and circular boundaries, can be programmed with ease, while more complex parts, such as cam contours (discussed in chapter 11) will require a working knowledge of analytical geometry.

Once the calculations are done, the programmer must combine the calculated data with the appropriate codes, acceptable to the respective CNC system, to produce a working part program.

The programmer's task is difficult, but possible. This task becomes harder with full three-axis contouring, and impossible for four- and five-axis applications.

Computer-assisted part programming offers a number of significant benefits. The simplicity of the language allows the programmer to describe the part contour with ease. The speed and accuracy of the computer produce tapes with significantly less lead time, without errors. Most computer-assisted systems allow the programmer to plot part geometry and the tool path with significant accuracy, thus cutting down on expensive machine time requirements for tape tryouts. The computer routinely transforms the calculated slide motions to the correct tape format, thus allowing the programmer to concentrate on the program, not on the features of the system on which the part is to be machined.

The benefits of computer-assisted programming are most significant in terms of programming time, accuracy, and flexibility, which can transform a CNC operation into a money-making concern.

12.1 COMPUTERIZED SYSTEMS FOR PART PROGRAMMING

The development of computer-assisted programming languages started in parallel with the production of the manufacturing systems. The manufacturers of NC systems, as well as the first users, realized that three-, four-, and five-axis systems can only be utilized to their fullest capabilities if a workable programming language is available to the users. This language had to be simple enough for the users to learn, and at the same time complex enough to address all the capabilities of the NC system. The first such language, called Automatically Programmed Tools (APT) was developed by MIT and became the most powerful full five-axis programming language. On behalf of the participating organizations, the APT long-range program is administered by the Illinois Institute of Technology Research Institution (IITRI). Additional capabilities are designed to meet new manufacturing needs, and continued maintenance and updating are assured. IITRI also promotes technical interchange among the community of users.

In parallel with the growth of the NC industry, numerous less complex computer-assisted languages were developed. Depending on the complexity of parts, as well as many other criteria, the users have a wide range of languages to choose from. Most of these share a general system concept. All computer-assisted NC programming languages use "English-like" words to describe the part, tool, type of operation, and the required auxilliary operations for the machining process. In some cases, the words may be abbreviated, sometimes to single letters. The NC program, using the programmed data, performs the mathematical calculations and compiles a "cutter location" list or file. The processed data is then recycled through another NC program called "postprocessor" or "link." The function of the latter is to change the processed data into coded tape formats acceptable to a specific machine-control system.

The NC program is universal. It is the same for all the different NC machines, which certainly eases the task of the programmer. The postprocessor, on the other hand, is a specific, special-purpose program, written for a well-defined machine-control system. The postprocessor knows all the programmable codes (G, M, T, S, etc.), acceleration and deceleration characteristics, ASCII or EIA tape formats, in addition to speeds, feeds, and physical limitations of slide motions. It is a "link" between the NC computer and the NC machine. A schematic representation of an NC computer-assisted process is illustrated in Fig. 12-1.

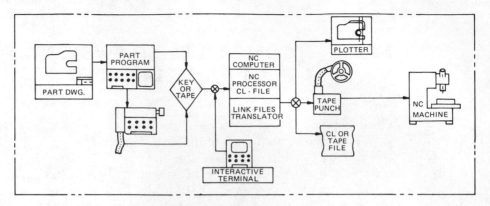

FIG. 12-1. NC computer system.

The figure illustrates a typical in-house system. The handwritten part program (manuscript) may be entered into the machine control unit (MCU) through a CRT or a teletype terminal. Alternatively, the program may be punched on tape and read into the MCU using a high-speed tape reader. Once the program has been entered, it must be edited for errors. The process is performed by the programmer through an interactive terminal. The programmer should obtain a "list" or "CL print" as well as a tape file printout prior to having the machine tape punched out. If the system is equipped with a plotter, a part outline and cutter center line path should be plotted.

The programmer must check all these data for accuracy and validity, make the final corrections where necessary, and finally punch out the machine control tape.

There are so many computer-assisted NC languages on the market that no purpose would be served in listing even the most common ones. New languages are developed constantly, and existing ones are continuously changed and improved. Although most existing languages were written for a specific computer, the newer ones can run on several different systems by design. The language incompatibility between the many different systems may present a problem from a teaching point of view. In practice, as a company normally uses only one such language—regardless of the number of different NC systems they may operate—this incompatibility does not pose an insurmountable problem.

12.2   SELECTING A COMPUTER-ASSISTED LANGUAGE

No company or organization should purchase a system outright without a thorough study of existing languages.

Numerous computer languages are also available on a time-sharing basis, an approach which allows the user to take advantage of modern computer facilities without large outlays to purchase and update the hardware. At the same time, the company can gain programming experience and find out about system limitations for their specific applications.

Most systems offer different levels of programming and processing capability. Only those that can be fully used should be selected, with written assurances, however, of advanced system capability and availability, if and when required, as well as costs of hardware and software, and implementation of any advanced processing capability.

One should examine the cost and availability of postprocessors or links. A long-range plan is necessary because changing an NC programming system to another one may be a very costly experience. A switch while one is on time-sharing is less costly, unless the company is committed to a long-term lease. Relocation of postprocessors and files may present problems, and retraining the programmers for another language will be time-consuming.

12.3   THE PROGRAMMING PROCESS

The process of programming consists of logical steps. The programmer must first define the coordinate system in which the NC program will be written, then define, in terms of computer statements, the part geometry. The type and size of tool as well as the tool motion instructions should be based on the process for achieving the part geometry.

Using the above information, the NC computer will perform all the necessary calculations for the postprocessor to code the NC tape. The postprocessor (link) will prepare a complete tape file including codes required to punch the tape, inch/metric dimensioning identification, ASCII (ISO) or EIA tape coding; motion codes for linear or circular interpolation (G01, G02, G03); canned cycle commands for drilling, boring, tapping, etc.; feeds, speeds, tool changes, and other auxilliary codes for setting the spindle, coolant, clamping, tailstock, etc.

12.4   THE PART GEOMETRY

The objective is to introduce the reader to basic two-dimensional programming, and the definitions will be limited to simple geometry. In order to show the similarity between the various computer languages, APT and COMPACT II have been selected for the subsequent examples.

12.4.1   The Point

The point is the smallest element of geometry and it is illustrated by examples in both languages selected in Fig. 12-2a and 12-2b.

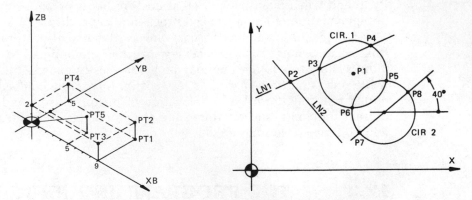

FIG. 12-2. Points defined by the intersection of:

a) Lines   **b) Circles.**

Point Defined by Coordinates

Compact II

```
DPT1,9XB,5YB,Zb
DPT2,9XB,5YB,2ZB
DPT3,9XB,YB,2ZB
DPT4,XB,5YB,2ZB
DPT5,5XB,3YB,2ZB
```

Any point can be defined in terms of the base coordinate system by the following general format:

DPTi,aXB,bYB,cZB

APT

P1 = POINT/9,5,0
P2 = POINT/9,5,2
P3 = POINT/9,0,2
P4 = POINT/0,5,2
P5 = POINT/5,3,2

and the general format of the APT definition is:

Pi = POINT/X,Y,Z

Point Defined by the Intersection of Lines and Circles

Assuming all the points to be located in the same X-Y plane, the points shown can be defined as follows:

Compact II

DPT1,CIR1/CNTR
DPT2,LN1,LN2
DPT3,LN1,CIR1,XS
DPT4,LN1,CIR1,YL
DPT5,CIR1,CIR2,XL
DPT6,CIR1,CIR2,XS
DPT7,CIR2,40CCW
DPT8,CIR2,140CW

The general definitions, corresponding to the Compact II examples, are shown below:

DefinePoinTi,symbol for circle/CeNTeR

DefinePoinTi,symbol for line,symbol for line

DefinePoinTi,symbol for line,symbol for circle,XLarge
                                                                XSmall
                                                                YLarge
                                                                YSmall

DefinePoinTi,symbol for circle,symbol for circle,XLarge
                                                                  XSmall
                                                                  YLarge
                                                                  YSmall

DefinePoinTi,symbol for circle,angle and direction of rotation.

APT

P1 = POINT/CENTER,CIR1

P2 = POINT/INTOF,LN1,LN2

P3 = POINT/XSMALL,INTOF,LIN1,CIR1

P4 = POINT/YLARGE,INTOF,LIN1,CIR1

P5 = POINT/XLARGE,INTOF,CIR1,CIR2

P6 = POINT/XSMALL,INTOF,CIR1,CIR2

P7 = POINT/CIR2,ATANGL,220   or P7 = POINT/CIR2,ATANGL,-140

P8 = POINT/CIR2,ATANGL,40

We are assuming here that the lines and circles used in the definitions of the various points have been defined previously in the same program. The general definitions equivalent to the above specific examples are given in the following APT geometry equations:

Pi = POINT/CENTER,symbol for circle

Pi = POINT/INTersectionOF,symbol for first line,symbol for the other line

Pi = POINT/XLARGE
           XSMALL
           YLARGE
           YSMALL,INTersectionOF,symbol for line,symbol for circle

Pi = POINT/XLARGE
           XSMALL
           YLARGE
           YSMALL, INTersectionOF,symbol for first circle, symbol for
           the other circle

Pi = POINT/Symbol for circle,ATanANGLe,angular value

There are many more ways of defining points in both APT and Compact II. Interested readers should consult the geometry sections of the appropriate programming manuals.

12.4.2   The Line

The computer program defines the line as an element of the part geometry. Mathematically, the defined line continues on beyond the part surface, in both directions, to "infinity."

Figure 12-3a and 12-3b will show simple forms of lines most common in the definition of part geometries.

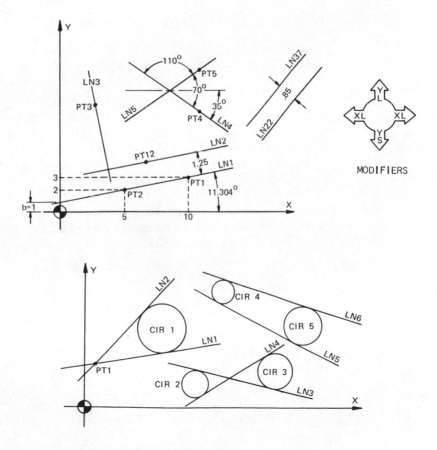

FIG. 12-3. Lines defined by:
   a) Coordinates, points, and angles.
   b) Using points and circles.

Line Definitions by Coordinates, Points, and Angles.

Compact II

DLN1,PT1,PT2
DLN2,LN1/1.25YL
DLN22,LN37/0.85XL
DLN2,PT12,PARLN1
DLN3,PT3,PERLN2
DLN4,PT4,PARX/35CW
DLN5,LN4,ROTXY-70 or DLN5,PT5,PARX/35CCW

The general definitions are shown below:

DefineLiNei,symbol for point,symbol for point
DefineLiNei,symbol for line,dimension parallel,modifier
DLNi,symbol for point,PARallel symbol for line
DLNi,symbol for point,PERpendicular symbol for line
DLNi,symbol for point,PARallel symbol for coordinate/dimension and symbol for rotation
DLNi,symbol for line,ROTation symbol for axes with direction and magnitude for rotation.

The same lines are again defined in APT. The wording and format are different, however, the similarity will be there.

APT

L1 = LINE/5,2,0,10,3,0   or   L1 = LINE/PT1,PT2

The line L1 can also be defined using the slope-intercept equation of the line, in its format

$$y = mx + b$$

The slope of the line, $m = \dfrac{y_1 - y_2}{x_1 - x_2} = \dfrac{3 - 2}{10 - 5} = \dfrac{1}{5} = 0.2$

The "y-intercept" is 1.0 inch and line 1 can now be written:

L1 = LINE/SLOPE,0.2,INTERC,1.0

Continuing with the rest of the geometry,

```
L2 = LINE/PARLEL,L1,YLARGE,1.25
L2 = LINE/P2,ATANGL,0,L1
L2 = LINE/P2,PARLEL,L1
```

Line 2 (L2) has been defined in three different ways using different modifiers. In the first case, a parallel condition to a previously defined line (L1) was used, modified in the "Y-large" direction.

In the second case we have used a previously defined line to indicate the direction, by stating a zero (0) angle condition, which is in fact a statement of parallelism. This definition allows the programmer to call for any angular relationship with the first line, by simply changing the angle zero (0) to whatever angle may be called in the drawing. P2 defined a fixed point through which our defined line crosses, in the part coordinate system.

In the third case, the word "PARLEL" (i.e., parallel) relates the direction of L1 and the location of P2.

There is no confusion in the selection of symbols when the same line is designated L2 or LN2, for instance. It is strictly a case of attempting to show two different languages using the same sketch. The first symbol is preferred in APT-like languages, while the latter is used in Compact II. Using them interchangeably within the same program would of course lead to an error message.

```
L3 = LINE/P3,ATANGL,90,L2
L3 = LINE/P3,PERPTO,L2
L4 = LINE/P4,ATANGL,110,L5
L4 = LINE/P4,ATANGL,-70,L5
L5 = LINE/P5,ATANGL,70,L4
L5 = LINE/P5,ATANGL,-110,L4
L22 = LINE/PARLEL,L37,YLARGE,0.85
```

The programming format for the L22 statement is:

```
Li = LINE/PARLEL, symbol for line,XLARGE
                                  XSMALL
                                  YLARGE
                                  YSMALL, magnitude of offset.
```

Line Definitions by Previously Defined Circles and Points.

Compact II

```
DLN1,PT1,CIR1,YS
DLN2,PT1,CIR1,YL
DLN3,CIR2,YL,CIR3,CROSS
DLN4,CIR2,YS,CIR3,CROSS
DLN5,CIR4,YS,CIR5
DLN6,CIR4,YL,CIR5
```

APT

```
L1 = LINE/P1,RIGHT,TANTO,C1
L2 = LINE/P1,LEFT,TANTO,C1
L3 = LINE/LEFT,TANTO,C3,RIGHT,TANTO,C2
L4 = LINE/LEFT,TANTO,C3,RIGHT,TANTO,C2
L5 = LINE/RIGHT,TANTO,C4,RIGHT,TANTO,C5
L6 = LINE/LEFT,TANTO,C4,LEFT,TANTO,C5
```

12.4.3   The Circle

The circle is the third most common geometric shape, after the point and the line. In the geometry section, the programmer defines the circle in its entirety. In the tool motion section, the program will outline which portion of this circle will be machined as part of the component boundary. Fig. 12-4 shows some of the more common definitions.

Compact II

```
DCIR1,P1,0.5R
DCIR1,PT2,PT3,PT4
```

Circle 1 is first defined through the location of its center and the magnitude of its radius, while in the second statement it is defined through three points on its circumference.

```
DCIR2,LN1/0.3YS,LN2/0.3XL,0.3R
```

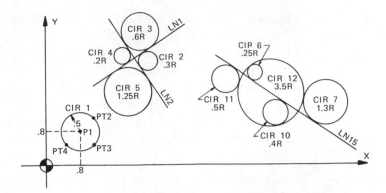

FIG. 12-4. Circle definitions.

This rather involved statement defines the circle as follows: The center is located at the intersection of two lines, one offset by 0.3 inch below line 1 and the other offset by the same amount to the right of line 2. The value of the circle radius is eventually defined as 0.3 inch.

```
DCIR3,LN1/0.6YL,LN2/0.6XL,0.6R
DCIR4,LN1/0.2XS,LN2/0.2YS,0.2R
DCIR5,LN1/1.25YS,LN2/1.25XS,1.25R
DCIR7,LN15/1.3YL,CIR12/1.3XL,1.3R
DCIR11,LN15/0.5YS,CIR12/0.5XS,0.5R
```

APT

As mentioned before, the APT program will accept any label for a given geometry. CIR1 or C1 will readily remind the programmer that circle 1 was defined, but so long as a label selected is not part of a list of "forbidden" terms, used by the processor itself and therefore potentially leading to confusion, any symbol may be selected.

```
CIR1 = CIRCLE/0.8,0.8,0,0.5
CIR1 = CIRCLE/CENTER,P1,RADIUS,0.5
```

In the first definition, the first three coordinates locate the center of the circle in X, Y, and Z, while the last dimension is the circle radius. The second definition achieves the same objective in a more obvious fashion, using the predefined label of point 1 where the center of the circle is located.

CIR2 = CIRCLE/YLARGE,L2,XLARGE,L1,RADIUS,0.3
CIR3 = CIRCLE/XLARGE,L2,YLARGE,L1,RADIUS,0.6
CIR4 = CIRCLE/XSMALL,L2,YLARGE,L1,RADIUS,0.2
CIR5 = CIRCLE/XSMALL,L2,YSMALL,L1,RADIUS,1.25
CIR7 = CIRCLE/YLARGE,L15,XLARGE,OUT,CIR12,RADIUS,1.3
CIR6 = CIRCLE/YLARGE,L15,XSMALL,IN,CIR12,RADIUS,0.25
CIR11 = CIRCLE/YSMALL,L15,XSMALL,OUT,CIR12,RADIUS,0.5
CIR10 = CIRCLE/YSMALL,L15,YSMALL,IN,CIR12,RADIUS,0.4

In both the APT and Compact II languages, there are a great deal more definitions for points, lines, and circles. In addition, more complex geometric solutions are available for vectors, planes, cones, ellipses, general quadratic surfaces which include hyperbolic paraboloids, ellipsoids, hyperboloids, etc.

It should also be mentioned that most "computer" words may be shortened to a single letter if required, thus decreasing the length of the program and the time required to write it.

Further in-depth discussion on the topics of geometry definition are beyond the objectives of this book. Interested readers are advised to obtain the programming manual corresponding to the language of their choice.

12.5   TOOL OR CUTTER STATEMENTS

Tool or cutter statements must be defined prior to any motion statements. The part geometry, described in the geometry definition section, is written using the dimensions of the component. Any tool motion without having specified a tool diameter or radius would create a tool path for a zero diameter tool. Specifying a cutter diameter will alter the tool path achieving cutter diameter compensation.

The tool statements will be discussed in more detail in the sample parts described in this chapter, as there are significant differences between milling and turning programming applications.

12.6   MOTION STATEMENTS

Motion statements are often called machine control instructions. All NC computer-assisted languages provide the programmer with a variety of statement formats for describing tool or machine motions, usually limited to the part geometry specified in the program. Other machine functions associated with tool movements, such as acceleration, deceleration, dwell, home, etc., are often interactive with a specific postprocessor or link. It also has to be pointed out that formats and wording statements cannot be changed and are to be used exactly as prescribed in the respective programming manuals. Manuscripts should be checked for accuracy and validity prior to entering the program into the computer. Most programs can be corrected off-line. On-line corrections should only be performed by experienced programmers as computer time is expensive.

12.7 COMPACT II MILLING—SAMPLE PROGRAMS

In the following pages, we shall present a number of sample programs for milling and turning applications, including explanations wherever required for clarification. These sample programs were compiled on a Compact II Series I in-house system, including a Datagen computer and a Hewlett-Packard plotter. The sample parts were machined on a Minimatic 500 horizontal machining center with a Fanuc 5M control, and a 4NE 12-position rear turret turning center with a Fanuc 6T control.

Example

Write a Compact II part program for the outside contour milling of the sample part illustrated in Fig. 12-5. The program should include both rough and finish milling.

12.7.1 Programming Data

Specific information must be available to the programmer prior to the start of the program. This includes cutter position at start, cutter diameter and length, setup data, including the name of the link (postprocessor). The following is a detailed listing.

The Name of the Link

The name of the link must be listed as the first statement of the program.

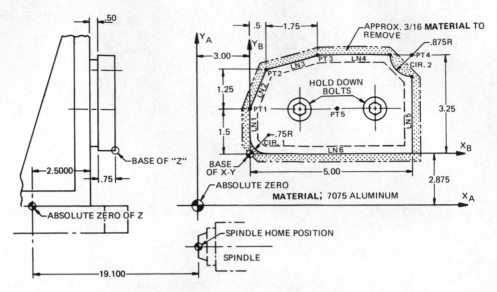

FIG. 12-5. Outside contour milling.

MACHIN,SEIKIFAN2

The name, "SEIKIFAN2," represents a specific machine and control combination.

Identification

Identification is the second statement of the program, and its contents will be punched out in tape in readable hole patterns in front of the machine control tape.

IDENT,SAMPLE PART,PARTNO.12345

Initialization

Initialization is the third statement in the program, and it will set the input and output modes for dimensional computation.

INIT,INCH/IN,INCH/OUT

The format shown above indicates that the part dimensions and tool sizes and motions will be given in inches, as will be the motions output in the machine control tape. Other possibilities of the above statement are:

> INIT,INCH/IN,METRIC/OUT
> INIT,METRIC/IN,INCH/OUT and
> INIT,METRIC/IN,METRIC/OUT

Only one initialization may be used in the NC program.

Setup

Setup is the fourth statement in the program. It is used to establish travel limits for the tool, specify the home position (where the tool changes will take place, normally) and the machine absolute zero. Additional information may be included in the setup to describe special link requirements for specific systems.

> SETUP,CMOD/1/2,MOD500/2,ABSO1,PALLET,LX,LY,12.50LZ,RPM3150

The first two statements after the word "SETUP" are link-related identifiers of specific equipment within a family of systems.

ABSO1 instructs the NC computer to calculate all tool motions in absolute mode. It will also generate a tape block to "zero" the control at its absolute zero position:

> N G92 X0 Y0 Z0

This statement will cause the control positional registers to show zero. As the machine will have been "zeroed" prior to running the tape, what takes place may be described as a synchronization of the machine and the control.

PALLET will generate a tape block for pallet transfer prior to any tool motion that takes place. This machine has two pallets, one for machining, the other for loading-unloading. The corresponding tape block is:

> N M60

nLX represents the distance along the axis from absolute zero (i.e., machine origin) to home position. As in our example $n = 0$, there is no distance between the above locations, and tool change will take place at machine origin. nLY is the distance along the Y-axis from absolute zero to home position. In our case, this $n = 0$ as well.

nLZ is the distance along the Z-axis from absolute zero to home position. In our example, this dimension is 12.50 inches.

RPMN is the spindle speed word, limiting the maximum programmable spindle speed at 3150 rpm. It also indicates to the control that on this system, the machine rpm may be programmed directly, without requiring M codes for various gear ranges.

Base

Base is the fifth statement in the program. The function of the base statement is to establish a base or coordinate system on the part for easier programming. In many cases, it coincides with the drawing datum point. The base is also used as a start point in the program.

In our example, the base is established at the front-left-top-corner of the part, using the statement:

    BASE,3XA,2.875YA,3.75ZA

The base has now been located along the three axes from the machine origin ("machine absolute zero").

We can now proceed with the problem solution in the following section.

12.7.2   Solution

We have seen the "start-up" portion of the program. We are relisting it below:

    MACHIN,SEIKIFAN2
    IDENT,SAMPLE,PART,PARTNO.12345
    INIT,INCH/IN,INCH/OUT
    SETUP,CMOD5/1/2,MOD500/2,ABSO1,PALLET,LX,LY,12.50LZ,RPM3150
    BASE,3XA,2.875YA,3.75ZA

We will now establish and write the part geometry statements.

Points:

```
DPT1,XB,1.5YB,ZB
DPT2,0.5XB,2.75YB,ZB
DPT3,2.25XB,3.25YB,ZB
DPT4,5.0XB,3.25YB,ZB
DPT5,2.5XB,1.625YB,ZB
```

Every time a point is written in the program, it should also be labeled on the programmer's drawing. This is required for subsequent ease of clarification, program checking, and debugging. Point 5 will be used to locate the part plot on the plotter, and it will be used in conjunction with a "draw" statement.

The reader has noticed that all the above points have been defined in the part coordinate system (XB,YB, and ZB). The Compact II language, however, allows the programmer to use the machine coordinate system as well, or any combination of the two, or another base if correctly defined. To illustrate this important capability, the definition of point 1 will be rewritten below using all possible combinations allowed so far by the system:

```
DPT1,3XA,4.375YA,3.75ZA
DPT1,3XA,1.5YB,ZB
DPT1,3XA,4.375YA,ZB
DPT1,XB,1.5YB,3.75ZA
DPT1,XB,4.375YA,3.75ZA
```

Obviously the programmer should only use one of these statements in the program. The reader should not, however, in spite of the freedom of the language, take advantage of mixed mode capabilities. This complicates the program and makes corrections more difficult. A change in the base statement would also require the programmer to change every statement that contains mixed mode definitions.

Lines

```
DLN1,XB
DLN2,PT1,PT2
DLN3,PT2,PT3
```

DLN4,3.25YB
DLN5,5XB
DLN6,YB

Circles

DCIR1,LN6/0.75YL,LN1/0.75XL,0.75R
DCIR2,PT4,0.875R

Having completed the geometry section, we shall now program a "draw" statement for the plotting of the part geometry and tool path.

DRAW,SCALE1,PT5,CNTR

The scale statement allows the programmer to increase or decrease the size of the plot in relation to the dimensions of the drawing.

e.g., SCALE0.5 will result in a half-size plot
SCALE2 will provide us with a plot twice the real part size.
PT5,CNTR instructs the computer to locate the center of the plot at point 5.

Tool Change

Tool change contains all the information relevant to machining and as such must precede the first motion statement in the program. Machines equipped with an automatic tool changer (ATC) are programmed using the word ATCHG. This must be programmed every time a tool change is required. In addition, this word initiates a number of other events, such as the axes being moved to the tool change point, the spindle and coolant being turned off, the spindle being oriented, the tool magazine being indexed, the spindle speed being selected, the spindle being started, the feed rate being set, etc.

ATCHG,TOOL1,5.75GL,0.500TD,800RPM,8IPM,NOX,NOY,CON,0.05STK
where TOOL1 will generate the appropriate code for the tool change. Machines equipped with random tool changers will index to position 1 and change,

while sequential tool changers will select the next available tool in the turret as tool No. 1.

nGL specifies the preset length of the tool. As shown in Fig. 12-6, n = 5.75 inches.

nTD is the tool diameter, necessary to the program for calculating the tool path. In our program n = 0.5 inches.

nRPM specifies the spindle speed in revolutions per minute. It will generate the following tape block:

N S800 M03

nIPM indicates the feed rate. In the program, n = 8 ipm.

NOX,NOY normally specify that no X-motion and no Y-motion are to take place, i.e., the tool is not to go "home" to its prescribed tool change location. As in our case the programmed tool change takes place at machine absolute zero, these words are not really required; they are in strictly for training purposes.

CON, meaning coolant on, will generate a tape block to turn the coolant on after the tool change has been completed.

nSTK specifies the thickness of material to be left for finishing, as "n" inches, in our case, 0.05. This amount will be added by the program to the cutter radius for compensation.

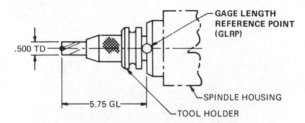

FIG. 12-6. Tool gage length.

There are additional features that can be programmed in the tool change instructions for specific purposes. As for the rest of the details, this information is available in the specific programming manual.

Motion Statements

The programmer must use previously defined part geometry to specify the desired motion pattern for the machine. The motion statements are many and varied, therefore, specific statements will be explained as used in the program. For our part, two passes have been specified around the part surface. The roughing pass will leave 0.050 inch stock on the outside contour, which will be removed by the finishing pass. To avoid writing the program twice, we have the capability of numbering the first and last motion statement in the machining sequence. Once the roughing cut has been programmed, the finishing can be programmed by only using a "Do" statement as shown below.

⟨1⟩MOVE, TOLN1/XS, TOLN6/0.3YS,0.1ZB

This statement will move the tool to point A, as illustrated in Fig. 12-7.

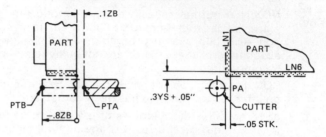

FIG. 12-7. Clearance and stock.

The word "Move" generates rapid tool motion, equivalent to the manual instruction

N . . . . . G00 Y . . . Y . . . Z . . .

⟨1⟩ represents statement number 1.

CUT,0.8ZB

This instruction will move the tool to point B.

The word "Cut" will generate tool motion in controlled feed rate. The amount of the feed rate was programmed in the "ATCHG" statement as 8 ipm.

CUT, PARLN1, PASTLN2

This statement, read as "cut, parallel to line 1 until the cutter gets completely on the other side of line 2" will bring the tool to point C, as shown in Fig. 12-8

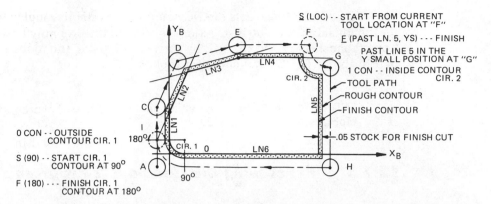

FIG. 12-8. Programmer's sketch for outside contour milling.

CUT, PARLN2, PASTLN3	Tool to point D
CUT, PARLN3, PASTLN4	Tool to point E
CUT, PARLN4, INCIR2, XS	Tool to point F
ICON, CIR2, CCW, S(LOC), F(PASTLN5,YS), 0.875R	Tool to point G
CUT, PARLN5, PASTLN6	Tool to point H
⟨2⟩OCON, CIR1, CW, S(90),F(180)	Tool to point I
ATCHG, TOOL2, 6.OGL, 0.5TD, 140 FPM, 0.005IPR, CON,NOX,NOY	

In the second tool change, above, different words were used to specify tool feed and speed.

nFPM specifies the tool surface velocity as n = 140 feet per minute. The NC computer will calculate the required spindle (tool) rpm using the equation

$$n = 12 \cdot v/\pi \cdot d$$

nIPR specifies the cutting feed as n = 0.005 inches per tool revolution.

S(LOC) means start from the location of the tool at the time the statement is read, on our part at point F.

F(PASTLN5,YS) indicates that the finish of the circular interpolation motion will see the cutter past line 5. There are two possibilities for this definition, after a 90° and a 270° angle. The first position is at G, below the latter, on the Y-axis, hence YSmall.

ICON is short for inside contour, in our case, that of circle 2.

OCON represents the outside contour of circle 1.

S(90), respectively, F(180) represent a start of the contour at 90° and an end at 180°, measured clockwise from the positive X-axis which is always 0° in Compact II.

D01/2,0 STK

This DO statement instructs the NC computer to establish a new tool path for the finishing cut. The instructions located in the program between motion statements ⟨1⟩ and ⟨2⟩ will be carried out, using the tool data specified in the "ATCHG" statement for tool No. 2.

END will terminate the program by generating an end of program tape block, as N M30

12.7.3   *Processing the Compact II Program*

This program, once entered into the disk file, was processed by the Compact II processor in the substitute run command mode. During the interactive process, the computer was instructed to create a "list file" and a "tape file."

The list file, along with the tool path and part geometry plots, as shown in Fig. 12-9, allows the programmer to check the program output and make changes as required. Once the programmer is satisfied that the program is correct, the computer is instructed to punch the machine control tape (off the tape file) and print its content. As instructed by the programmer, the machine control tape will be punched out in EIA or ASCII (ISO) format.

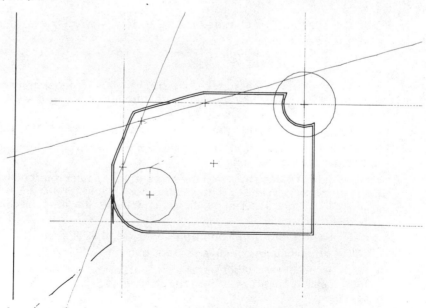

FIG. 12-9. Computer produced plot for outside contour milling.

Source Program

```
+P
PUNCHING TAPE TOO?
IDENT?
↑   ↑>HHH><BBB$↑↑BB↑BB↑ ↑↑JJBB4JJJ$↑JJBBBB↑BB↑ DBB↑BB↑HH@@>HHH>↑ ↑&BJR"B█
MACHINE,SEIKIFAN2
IDENT,SAMPLE PART,PART NO.12345
INIT,INCH/IN,INCH/OUT
SETUP,CMOD5/1/2,MOD500/2,ABS01,PALLET,LX,LY,LZ,RPM3150
BASE,3XA,2.875YA,3.75ZA
DPT1,XB,1.5YB,ZB
DPT2,.5XB,2.75YB,ZB
DPT3,2.25XB,3.25YB,ZB
DPT4,5.0XB,3.25YB,ZB
DPT5,2.5XB,1.625YB,ZB
DLN1,XB
DLN2,PT1,PT2
DLN3,PT2,PT3
DLN4,3.25YB
DLN5,5.0XB
DLN6,YB
DCIR1,LN6/.75YL,LN1/.75XL,.75R
DCIR2,PT4,.875R
DRAW,SCALE1,PT5,CNTR
ATCHG,TOOL1,5.75GL,.500TD,800RPM,8IPM,NOX,NOY,CON,.05STK
<1>MOVE,TOLN1/XS,TOLN6/.3YS,.1ZB
CUT,.8Z
CUT,PARLN1,PASTLN2
CUT,PARLN2,PASTLN3
CUT,PARLN3,PASTLN4
CUT,PARLN4,INCIR2,XS
ICON,CIR2,CCW,S(LOC),F(PASTLN5,YS),.875R
CUT,PARLN5,PASTLN6
<2>OCON,CIR1,CW,S(90),F(180)
ATCHG,TOOL2,6.0GL,.5TD,140FPM,.005IPR,CON,NOX,NOY
DO1/2,0STK

END
```

List File

```
>MACHINE,SEIKIFAN2
MAIN 60980 LINK 21580 SYS 42578 L# 1483

>IDENT,SAMPLE PART,PART NO.12345
00-00-00  00:00

>INIT,INCH/IN,INCH/OUT

>SETUP,CMOD5/1/2,MOD500/2,ABS01,PALLET,LX,LY,LZ,RPM3150

>BASE,3XA,2.875YA,3.75ZA
= X 3. Y 2.875 Z 3.75

>DPT1,XB,1.5YB,ZB
= X 3. Y 4.375 Z 3.75

>DPT2,.5XB,2.75YB,ZB
= X 3.5 Y 5.625 Z 3.75
```

```
>DPT3,2.25XB,3.25YB,ZB
= X 5.25 Y 6.125 Z 3.75

>DPT4,5.0XB,3.25YB,ZB
= X 8. Y 6.125 Z 3.75

>DPT5,2.5XB,1.625YB,ZB
= X 5.5 Y 4.5 Z 3.75

>DLN1,XB
= X 3. Y . A 90.

>DLN2,PT1,PT2
= X 3. Y 4.375 A 68.1986

>DLN3,PT2,PT3
= X 3.5 Y 5.625 A 15.9454

>DLN4,3.25YB
= X . Y 6.125 A .

>DLN5,5.0XB
= X 8. Y . A 90.

>DLN6,YB
= X . Y 2.875 A .

>DCIR1,LN6/.75YL,LN1/.75XL,.75R
= X 3.75 Y 3.625 R .75

>DCIR2,PT4,.875R
= X 8. Y 6.125 R .875

>DRAW,SCALE1,PT5,CNTR

>ATCHG,TOOL1,5.75GL,.500TD,800RPM,8IPM,NOX,NOY,CON,.05STK
 N0000 M60
 N0010 G92 X0 Y0 Z0
 N0020 G91 G28 Z0
 N0030 G90 G00 X0 Y0
 N0040 M06
 N0050 S800 M03

><1>MOVE,TOLN1/.XS,TOLN6/.3YS,.1ZB
 N0060 G90 G00 Z-9.6000 M08
 N0070 X 2.7000 Y 2.2750

>CUT,.8Z
 N0080 G01 Z-10.4000 F8.00

>CUT,PARLN1,PASTLN2
 N0090 Y 4.4328

>CUT,PARLN2,PASTLN3
 N0100 X 3.2761 Y 5.8730

>CUT,PARLN3,PASTLN4
 N0110 X 5.2081 Y 6.4250

>CUT,PARLN4,INCIR2,XS
 N0120 X 7.5095
>ICON,CIR2,CCW,S(LOC),F(PASTLN5,YS),.875R
 N0130 G03 X 7.4250 Y 6.1250 I .4905 J-.3000 F5.58
 N0140 X 8.0000 Y 5.5500 I .5750
 N0150 X 8.3000 Y 5.6345 J .5750
```

```
> CUT, PARLN5, PASTLN6
  NO160 G01 Y 2.5750 F8.00

> <2>OCON, CIR1, CW, S(90), F(180)
  NO170 X 3.7500
  NO180 G02 X 2.7000 Y 3.6250 J 1.0500

> ATCHG, TOOL2, 6.OGL, .5TD, 140FPM, .005IPR, CON, NOX, NOY
  NO190 GOO
  NO200 ZO
  NO210 MO6

> DO1/2, OSTK
*DO
  NO220 S1069 MO3
  NO230 G90 GOO Z-9.8500 MO8
  NO240 X 2.7500 Y 2.8250
*DO
  NO250 G01 Z-10.6500 F5.35
*DO
  NO260 Y 4.4232
*DO
  NO270 X 3.3134 Y 5.8317
*DO
  NO280 X 5.2150 Y 6.3750
*DO
  NO290 X 7.4272
*DO
  NO300 G03 X 7.3750 Y 6.1250 I .5728 J-.2500 F3.82
  NO310 X 8.0000 Y 5.5000 I .6250
  NO320 X 8.2500 Y 5.5522 J .6250
*DO
  NO330 G01 Y 2.6250 F5.35
*DO
  NO340 X 3.7500
  NO350 G02 X 2.7500 Y 3.6250 J 1.0000

>

> END
  NO360 GOO XO YO
  NO370 ZO
  NO380 M30
END MIN: 6.2 FT: 14.5 MTR: 4.4

ERRORS DETECTED = 0
```

Tape File

```
TAPE FILE: /CNC-JP100T/
EIA?
NO
OUTPUT TO: TPT
TURN PUNCH ON, HIT CR
4JJJ$↑JJBBBB↑BB↑ DBB↑BB↑HH@@>HHH>↑ ↑&BJR"<BBB<<BBB<<BBB<<BBB<<BBB<<BBB<▪
```

%

```
NOOOOM60
NO010G92XOYOZO
NO020G91G28ZO
```

```
N0030G90G00X0Y0
N0040M06
N0050S800M03
N0060G90G00Z-96000M08
N0070X27000Y22750
N0080G01Z-104000F800
N0090Y44328
N0100X32761Y58730
N0110X52081Y64250
N0120X75095
N0130G03X74250Y61250I4905J-3000F558
N0140X80000Y55500I5750
N0150X83000Y56345J5750
N0160G01Y25750F800
N0170X37500

N0180G02X27000Y36250J10500
N0190G00
N0200Z0
N0210M06

N0220S1069M03
N0230G90G00Z98500M08
N0240X27500Y28250
N0250G01Z-106500F535
N0260Y44232
N0270X33134Y58317
N0280X52150Y63750
N0290X74272
N0300G03X73750Y61250I5728J-2500F382
N0310X80000Y55000I6250
N0320X82500Y55522J6250
N0330G01Y26250F535
N0340X37500
N0350G02X27500Y36250J10000
N0360G28X0Y0
N0370G28Z0
N0380M30
```

EXAMPLE

Write a Compact II program for outside contour milling and hole pattern drilling and tapping, for the part illustrated in Fig. 12-10. The program should leave 0.1 inch stock on the outside contour for finish milling.

Programming data:
Tool No. 1: 0.5-inch diameter, 6.1 GL (gage length), end mill
Tool No. 2: 0.125-inch diameter, 5.5 GL, center drill
Tool No. 3: 0.3125-inch diameter 5.75 GL, drill
Tool No. 4 : 0.375-inch diameter 5.25 GL, tap

The rest of the machine and link data will be identical to the previous example.

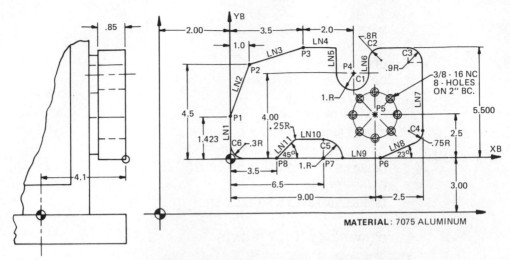

FIG. 12-10. Milling, drilling and tapping.

Solution:

Start-up:

        MACHIN, SEIKIFAN2
        IDENT, SAMPLE PART MILL-DRILL-TAP
        INIT, INCH/IN, INCH/OUT
        SETUP,CMOD/1/2,MOD500/2,ABSO1,PALLET,XL,YL,12.50LZ,RPM3150
        BASE,2XA,3YA,4.1ZA

Geometry:

        DPT1,XB,1.423YB,ZB
        DPT2,1.0XB,4.5YB,ZB
        DLN1,XB
        DLN2,PT1,PT2
        DPT3,3.5XB,5.5YB,ZB
        DLN3,PT2,PT3
        DLN4,5.5YB
        DLN5,4.5XB
        DLN6,6.5XB
        DLN7,11.5XB
        DPT4,5.5XB,4YB,ZB
        DCIR1,PT4,1.0R
        DPT5,9.0XB,2.5YB,ZB
        DPT6,9.0XB,YB,ZB

```
        DLN8,PT6,23CCW
        DLN9,YB
        DLN10,1.0YB
        DPT7,6.5XB,YB,ZB
        DPT8,3.5XB,YB,ZB
        DLN11,PT8,45CCW
        DCIR2,LN6/0.8XL,LN4/0.8YS,0.8R
        DCIR3,LN4/0.9YS,LN7/0.9XS,0.9R
        DCIR4,LN7/0.75XS,LN8/0.75YL,0.75R
        DCIR5,PT7,1.0R
        DCIR6,LN9/0.3YL,LN1/0.3XL,0.3R
```

Plot:
```
        DRAW,SCALE (1/2), PT5,CNTR
```
1st tool change for milling:
```
        ATCHG,TOOL1,6.1GL,0.5TD,850RPM,8.5IPM,0.1STK,CON
```

Rough milling:
```
        ⟨4⟩MOVE,OFFLN1/XS,OFFLN9/0.3YS,0.1ZB
        CUT, 0.9ZB
        CUT,PARLN1,OFFLN2/YL
        CUT,PARLN2,PASTLN3
        CUT,PARLN3,PASTLN4
        CUT,PARLN4,PASTLN5
        ICON,CIR1,CCW,S(180),F(0)
        OCON,CIR2,CW,S(180),F(270)
        OCON,CIR3,CW,S(270),F(0)
        OCON,CIR4,CW,S(0),F(TANLN8)
        CUT,PARLN8,OFFLN9/YS
        CUT,PARLN9,INCIR5,XL
        ICON,CIR5,CCW,S(0),F(90)
        CUT,PARLN10,OFFLN11//XL
        CUT,PARLN11,OFFLN9/YS
        ⟨5⟩OCON,CIR6,CW,S(90),F(180)
        DO 4/5,0 STK
```

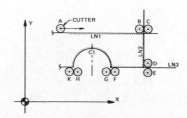

FIG. 12-11. Sketch illustrating various motion commands.

Interpretation of motion commands OFFLN*i*/XL and INsideCIRcle*i*,XL
/XS XS
/YL YL
/YS YS

Using Fig. 12-11,
from point A to point B: CUT,PARLN1,OFFLN2/XS
from point A to point C: CUT,PARLN1,OFFLN2/XL
from point C to point D: CUT,PARLN2,OFFLN3/YL
from point C to point E: CUT,PARLN2,OFFLN3/YS
from point E to point F: CUT,PARLN3,OUTCIR1,XL
from point E to point G: CUT,PARLN3,INCIR1,XL
from point E to point H: CUT,PARLN3,INCIR1,XS
from point E to point K: CUT,PARLN3,OUTCIR1,XS

Center drilling:

    ATCHG,TOOL2,5.5GL,0.125TD,2100RPM,4 IPM,NOX,NOY,CON,118TPA
    DRL,PT5,2.0BC,8EQSP,CW,S(0),0.3DP

where:

DRL is a major word that initiates the drill cycle

PT*i* specifies the number of the point where the bolt circle center is located

*i*EQSP represents the number of equally spaced holes, in our case *i* = 8

CW specifies the direction, clockwise, in which the holes will be successively drilled

S(*i*) is the starting angle, respectively the starting angular location of the first hole. In the program, *i* = 0 on the bolt circle

*n*BC represents the diameter $n - 2$ of the bolt circle

*i*DP specifies the depth of drilling from 0ZB, in our case, *i* = 0.3

*i*TPA specifies the tool point angle *i* = 118°

Drilling:

    ATCHG,TOOL3,5.75GL,0.3125TD,680RPM,5 IPM,NOX,NOY,CON,118TPA
    DRL,PT5,2.0BC,8EQSP,CW,S(0),0.85THRU

*i*THRU specifies that the holes are to be drilled through the part whose thickness *i* = 0.85 inches.

Tapping:

    ATCHG,TOOL4,5.25GL,0.375TD,(1/16)LEAD,100RPM,NOX,NOY,CON
    FLT,PT5,2.0BC,8EQSP,CW,S(0),0.85THRU

*i*LEAD represents the lead or pitch of the tapped hole; *i* = 1/16 = 0.0625 inch. The Compact II computer will use this value, in conjunction with the rpm to compute the tap feed.

    END

The list and tape file printouts are shown below, followed by Fig. 12-12 which illustrates the part geometry and tool path plot for the sample part.

Source Program

```
MACHINE,SEIKIFAN2
IDENT,MILL-DRILL-TAP
INIT,INCH/IN,INCH/OUT
SETUP,CMOD5/1/2,MOD500/2,ABS01..,PALLET,LX,LY,LZ,RPM3150
BASE,2XA,3YA,4.1ZA
DPT1,XB,1.423YB,ZB
DPT2,1.0XB,4.5YB,ZB
DLN1,XB
DLN2,PT1,PT2
DPT3,3.5XB,5.5YB,ZB
DLN3,PT2,PT3
DLN4,5.5YB
DLN5,4.5XB
DLN6,6.5XB
DLN7,11.5XB
DPT4,5.5XB,4YB,ZB
DPT5,9XB,2.5YB,ZB
DPT6,9.7XB,YB,ZB
DLN8,PT6,23CCW
DLN9,YB
DLN10,1YB
DCIR1,PT4,1R
DPT7,6.5XB,YB,ZB
DPT8,3.5XB,YB,ZB
DLN11,PT8,45CCW
DCIR2,LN6/.8XL,LN4/.8YS,.8R
DCIR3,LN4/.9YS,LN7/.9XS,.9R
DCIR4,LN7/.75XS,LN8/.75YL,.75R
DCIR5,PT7,1.0R
DCIR6,LN9/.3YL,LN1/.3XL,.3R
DRAW,SCALE(1/2),PT5,CNTR
ATCHG,TOOL1,6.1GL,.5TD,850RPM,8.5IPM,.1STK,CON
<1>MOVE,OFFLN1/XS,OFFLN9/.3YS,.1ZB
CUT,.9ZB
CUT,PARLN1,OFFLN2/YL
CUT,PARLN2,PASTLN3
CUT,PARLN3,PASTLN4
CUT,PARLN4,PASTLN5
ICON,CIR1,CCW,S(180),F(0)
OCON,CIR2,CW,S(180),F(270)
OCON,CIR3,CW,S(270),F(0)
OCON,CIR4,CW,S(0),F(TANLN8)
CUT,PARLN8,OFFLN9/YS
CUT,PARLN9,INCIR5,XL
ICON,CIR5,CCW,S(0),F(90)
CUT,PARLN10,OFFLN11/XL
CUT,PARLN11,OFFLN9/YS
<2>OCON,CIR6,CW,S(90),F(180)
ATCHG,TOOL2,5.5GL,.125TD,2100RPM,4IPM,NOX,NOY,CON,118TPA
DRL,PT5,2.0BC,8EQSP,CW,S(0),.3DP
ATCHG,TOOL3,5.75GL,.3125TD,680RPM,5IPM,NOX,NOY,CON,118TPA
DRL,PT5,2.0BC,8EQSP,CW,S(0),.85THRU
ATCHG,TOOL4,5.25GL,.375TD,(1/16)LEAD,100RPM,NOX,NOY,CON
FLT,PT5,2.BC,8EQSP,CW,S(0),.85THRU
END
```

+

List File

```
>MACHINE,SEIKIFAN2
MAIN 60980 LINK 21580 SYS 42578 L# 1483

>IDENT,MILL-DRILL-TAP
00-00-00  00:00

>INIT,INCH/IN,INCH/OUT

>SETUP,CMOD5/1/2,MOD500/2,ABS01.,PALLET,LX,LY,LZ,RPM3150

>BASE,2XA,3YA,4.1ZA
= X 2. Y 3. Z 4.1

>DPT1,XB,1.423YB,ZB
= X 2. Y 4.423 Z 4.1

>DPT2,1.0XB,4.5YB,ZB
= X 3. Y 7.5 Z 4.1

>DLN1,XB
= X 2. Y . A 90.

>DLN2,PT1,PT2
= X 2. Y 4.423 A 71.9963

>DPT3,3.5XB,5.5YB,ZB
= X 5.5 Y 8.5 Z 4.1

>DLN3,PT2,PT3
= X 3. Y 7.5 A 21.8014

>DLN4,5.5YB
= X . Y 8.5 A .

>DLN5,4.5XB
= X 6.5 Y . A 90.

>DLN6,6.5XB
= X 8.5 Y . A 90.

>DLN7,11.5XB
= X 13.5 Y . A 90.

>DPT4,5.5XB,4YB,ZB
= X 7.5 Y 7. Z 4.1

>DPT5,9XB,2.5YB,ZB
= X 11. Y 5.5 Z 4.1

>DPT6,9.7XB,YB,ZB
= X 11.7 Y 3. Z 4.1

>DLN8,PT6,23CCW
= X 11.7 Y 3. A 23.

>DLN9,YB
= X . Y 3. A .
```

```
>DLN10,1YB
= X . Y 4. A .

>DCIR1,PT4,1R
= X 7.5 Y 7. R 1.

>DPT7,6.5XB,YB,ZB
= X 8.5 Y 3. Z 4.1

>DPT8,3.5XB,YB,ZB
= X 5.5 Y 3. Z 4.1

>DLN11,PT8,45CCW
= X 5.5 Y 3. A 45.

>DCIR2,LN6/.8XL,LN4/.8YS,.8R
= X 9.3 Y 7.7 R .8

>DCIR3,LN4/.9YS,LN7/.9XS,.9R
= X 12.6 Y 7.6 R .9

>DCIR4,LN7/.75XS,LN8/.75YL,.75R
= X 12.75 Y 4.2605 R .75

>DCIR5,PT7,1.0R
= X 8.5 Y 3. R 1.

>DCIR6,LN9/.3YL,LN1/.3XL,.3R
= X 2.3 Y 3.3 R .3

>DRAW,SCALE(1/2),PT5,CNTR

>ATCHG,TOOL1,6.1GL,.5TD,850RPM,8.5IPM,.1STK,CON

><1>MOVE,OFFLN1/XS,OFFLN9/.3YS,.1ZB
ABS X 1.65 Y 2.35 Z 4.2

>CUT,.9ZB
ABS X 1.65 Y 2.35 Z 5.

>CUT,PARLN1,OFFLN2/YL
ABS X 1.65 Y 4.4785Z 5.

>CUT,PARLN2,PASTLN3
ABS X 2.7178Y 7.7641Z 5.

>CUT,PARLN3,PASTLN4
ABS X 5.4326Y 8.85 Z 5.

>CUT,PARLN4,PASTLN5
ABS X 6.85 Y 8.85 Z 5.

>

>

>ICON,CIR1,CCW,S(180),F(0)
ABS X 8.15 Y 7. Z 5.

>OCON,CIR2,CW,S(180),F(270)
ABS X 9.3 Y 8.85 Z 5.
```

```
>OCON,CIR3,CW,S(270),F(0)
ABS X 13.85 Y 7.6 Z 5.

>OCON,CIR4,CW,S(0),F(TANLN8)
ABS X 13.1798Y 3.2479Z 5.

>CUT,PARLN8,OFFLN9/YS
ABS X 11.7712Y 2.65 Z 5.

>CUT,PARLN9,INCIR5,XL
ABS X 9.0477Y 2.65 Z 5.

>ICON,CIR5,CCW,S(0),F(90)
ABS X 8.5 Y 3.65 Z 5.

>CUT,PARLN10,OFFLN11/XL
ABS X 6.645 Y 3.65 Z 5.

>CUT,PARLN11,OFFLN9/YS
ABS X 5.645 Y 2.65 Z 5.

><2>OCON,CIR6,CW,S(90),F(180)
ABS X 1.65 Y 3.3 Z 5.

>ATCHG,TOOL2,5.5GL,.125TD,2100RPM,4IPM,NOX,NOY,CON,118TPA

>DRL,PT5,2.0BC,8EQSP,CW,S(0),.3DP
ABS X 11.7071Y 6.2071Z 4.2

>ATCHG,TOOL3,5.75GL,.3125TD,680RPM,5IPM,NOX,NOY,CON,118TPA

>DRL,PT5,2.0BC,8EQSP,CW,S(0),.85THRU
ABS X 11.7071Y 6.2071Z 4.2

>ATCHG,TOOL4,5.25GL,.375TD,(1/16)LEAD,100RPM,NOX,NOY,CON

>FLT,PT5,2.BC,8EQSP,CW,S(0),.85THRU
ABS X 11.7071Y 6.2071Z 4.2

>END
END


+
```

Tape File

```
 P
&↑       SEIKIFAN2         00-00-00  00:00
ILL-DRILL-TAP    M
=
N0000M60
N0010G92X0Y0Z0
N0020G91G28Z0
N0030G90G00X0Y0
N0040M06
N0050S850M03
N0060G90G00Z-103000M08
N0070X16500Y23500
N0080G01Z-111000F850
N0090Y44785
N0100X27178Y77641
```

```
N0110X54326Y88500
N0120X68500
N0130Y70000
N0140G03X75000Y63500I6500F614
N0150X81500Y70000J6500
N0160G01Y77000F850
N0170G02X93000Y88500I11500
N0180G01X126000
N0190G02X138500Y76000J-12500
N0200G01Y42605
N0210G02X131798Y32479I-11000
N0220G01X117712Y26500
N0230X90477
N0240X91500Y30000
N0250G03X85000Y36500I-6500F614
N0260G01X66450F850
N0270X56450Y26500
N0280X23000
N0290G02X16500Y33000J6500
N0300G00
N0310Z0
N0320M06
N0330S2100M03
N0350G81X120000Y55000Z-92624R-97000F400
N0360X117071Y47929
N0370X110000Y45000
N0380X102929Y47929
N0390X100000Y55000
N0400X102929Y62071
N0410X110000Y65000
N0420X117071Y62071
N0430G80Z0
N0440M06

N0450S680M03
N0470G81X120000Y55000Z-88061R-99500F500
N0480X117071Y47929
N0490X110000Y45000
N0500X102929Y47929
N0510X100000Y55000
N0520X102929Y62071
N0530X110000Y65000
N0540X117071Y62071
N0550G80Z0
N0560M06

N0570S100M03
N0590G84X120000Y55000Z-84000R-94500F625
N0600X117071Y47929M03
N0610X110000Y45000M03
N0620X102929Y47929M03
N0630X100000Y55000M03
N0640X102929Y62071M03
N0650X110000Y65000M03
N0660X117071Y62071M03
N0670G80M03
N0680G28X0Y0
N0690G28Z0
N0700M30
*
```

The plot allows the programmer to check the part geometry (points, lines, circles, etc.) against the tool path with ease. Most simple plots use two colors, one for the geometry and the other for the tool path.

Plots of complex parts using numerous tools can be color-coded with a different color pen for each path. The NC computer, based on the program information, has completed all the necessary calculations for the tool path and has printed the information required in the list file. The link (postprocessor) received this calculated data and translated it into acceptable codes for the respective NC machine, including the appropriate circular interpolation codes (G02 and G03) with the correct dimensional words (X and Y) and unit vectors (I and J). It selected the appropriate codes and formats for drilling and tapping (G81 and G84). The reader can clearly appreciate that any comparison between manual and computer-assisted programming is weighed in favor of the computer.

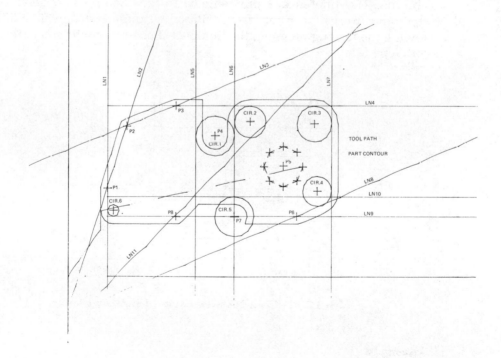

FIG. 12-12. Computer produced plot of part shown in Fig. 12-10.

12.8   COMPACT II TURNING—SAMPLE PROGRAM

This chapter on computer-assisted programming would not be complete without a turning sample program. The basic definitions are similar, in many cases identical to milling. However, there are some essential differences, derived from the concept of machining itself. In turning, the part is rotated and the tool moves in a two-dimensional plane to generate the prescribed contour. The motion statements are far simpler to analyze in the absence of a third axis.

The most popular CNC is the horizontal programmable tool changer type, best known in industry as the CNC turning center. These are built with "front" or "rear" turrets (the four-axis turning centers have both front and rear tool turrets, independently programmable for simultaneous motion in different directions).

The basic coordinate system of the machine is illustrated in Fig. 12-13. The part program is identical for both front or rear turret machining. The difference is programmed in the "Setup" statement using the minor words "rear" or "front." Based on these instruments and following the turning program, the Compact II computer will create the list and tape files.

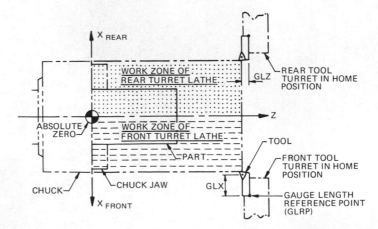

FIG. 12-13. Coordinate system of turning center.

Example

Write a Compact II part program for the finish turning of a typical part as illustrated in Fig. 12-14. The program will contain facing, contouring, grooving, threading, and drilling. As these operations are among the most

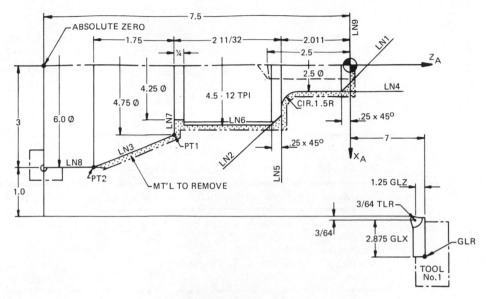

FIG. 12-14. Typical turning part.

common in part programs for turning centers, it is expected that they will provide a comprehensive overview of the Compact II programming process for turning.

12.8.1   Solution

1. Select the tool types and establish their dimensions as required for machining and programming.
    The Compact II tool statement requires us to specify the following dimensions:

TOOLn where n is the tool turret position

GLXn represents the gage length in the X-direction and n is the dimension from the Gage Length Reference Point (GLRP) to the center of the tool tip radius.

GLZn is the same requirement in the Z-direction

TLRn specifies the tool (nose) radius of value n.

The above requirements are written as follows:

ATCHG,TOOL1,OFFSET1,GLX2.875,GLZ1.25,TLR(3/64),0.008IPR,
400FPM,RANGE1

The minor word RANGEn, used in conjunction with our first tool for facing, is only defined in the first tool statement. Its function is to generate the appropriate M code for a gear change and in our program $n = 1$. The balance of the minor words have already been explained in the milling example.

The tooling data can best be summarized in Table 12-1.

TABLE 12-1 TOOLING DATA							
Operation	Tool No.	GLX	GLZ	TLR	Offset	IPR	FPM
Facing	1	2.875	1.25	3/64	1	0.008	400
Contouring	2	2.875	1.25	1/32	2	0.015	400
Grooving	3	2.5	1.0	0	3	0.004	80
Threading	4	2.75	1.0	0	4	—	50
Drilling	5	0	8.0	118TPA	5	0.006	80

2. Establish the part geometry by labeling points, lines, and circles on the part drawing, as shown in Fig. 12-12.

3. Establish the program base and label the coordinate system on the part drawing. In programming for turning centers, the XA should always be at $X = 0$, while the ZA may be located at any length from the absolute zero. The latter should, however, be located on the part surface, as illustrated in Fig. 12-12.

12.8.2   Writing the Part Program

Setup instructions.

    MACHIN,DEMOLATHE

where Demolathe is the name assigned to the link (postprocessor)

IDENT,CNC-JP-11 FRONT LATHE
SETUP,MOD2/1,CMOD2/2,ABSO2,ZEROS3,RPM 2500,X3 + 1 + (3/64) + 2.875,
(continued)
Z7.5 + 7,LIMIT(X0/10,Z2/18),FRONT

The X- and Z- dimensions are established by the programmer. These dimensions should be far enough from the part to allow the operator to load and unload the parts.

ABSOn will cause the tape output to be in absolute mode, reflecting the tool point motions. n = 2 will generate a tape block containing a G50 code after each tool change.

ZEROSn will generate a machine control tape in decimal point format when n = 3 as programmed.

LIMIT specifies the absolute travel limits to be checked in relationship to the primary turret GLRP.

```
INIT, INCH/IN,INCH/OUT
BASE,XA,7.5ZA
```

Part geometry:

```
DLN1,ZB,2.0D,45CCW
DLN2,4.5D,-2.011-0.25ZB,45CCW
```

The 4.5D,-2.011-0.25ZB defines a point. The computer will generate a line through this point, parallel to the Z-axis.

The 45CCW minor word will then rotate this same line 45° in the counterclockwise direction to form the required 45° chamfer.

```
DPT1,4.75D,-2.0-(11/32)-2.011ZB
DPT2,6D,-1.75-2.0-(11/32)-2.011ZB
DLN3,PT1,PT2
DLN4,2.5D
DLN5,2.011ZB
DLN6,4.5D
DLN7,-2.011-2.0-(11/32)ZB
DLN8,6.0D
DLN9,ZB
DCIR1,LN5/0.5ZL,LN4/0.5XL,0.5R
```

The reader has noticed, by now, the difference in the line definitions. The lines parallel to the Z-axis can be simply defined in terms of their diameter, right off the drawing.

Tool change and motion statements for facing:

```
ATCHG,TOOL1,OFFSET1,GLX2.875,GLZ1.25,TLR(3/64),0.008IPR,400FPM,RANGE1
MOVEC,OFFLN1/0.35XL,OFFLN9/ZL
```

The MOVEC major word (Move to Cut) generates a rapid traverse motion with automatic deceleration to feed rate before the programmed point has been reached.

```
CUT,ONLN(XB),CSS/ON
CUT,0.05X,0.05Z
```

ONLN(XB) will move the tool point parallel to line 9, onto the center line of the part.

CSS/ON, Constant surface speed on, will generate the appropriate G code to increase the spindle rpm as the tool moves inward to the center of the part. This instruction is normally programmed when the machined contour changes in diameter.

Tool change and motion statements for contour turning:

```
ATCHG,TOOL2,OFFSET2,GLX2.875,GLZ1.25,TLR(1/32),400FPM,0.015IPR
MOVEC,OFFLN1/XL,OFFLN9/0.1ZL
CUT,PARLN1,OFFLN4/XL,CSS/ON
ICON,CIR1,S(TANLN4),F(TANLN5),CCW
CUT,PARLN5,OFFLN2/XL
CUT,PARLN2,OFFLN6/XL
CUT,PARLN6,OFFLN7/ZL
CUT,PARLN7,OFFLN3/XL
CUT,PARLN3,OFFLN8/XL
MOVE,OFFLN9/0.1ZL
```

The above motion statements are easy to follow as they are directly related to the part geometry.

Tool change and motion statements for the turning of the groove:

ATCHG,TOOL3,OFFSET3,GLX2.5,GLZ1.0TLR,CCS/OFF,80FPM,RPMD4.25, (cont)
0.004IPR

This is yet another easy way of defining the spindle RPM in terms of the part diameter, 4.25, in conjunction with a surface speed of 80 FPM. The Compact II computer will substitute these values into the equation below:

$$n = \frac{v \cdot 12}{4.25 \cdot \pi} = 71.9 \text{ rpm}$$

This spindle speed is reflected in the second tape block following the tool change in the tape file:

N0040 G97 S0071 M40

It is seen that the computer truncated 71.9 rpm to 71 in the S word.

MOVE,OFFLN3/0.05XL,OFFLN7/ZL
CUT,4.25D
CUT,OFFLN3/0.05XL,PARLN7

The "move" statement will rapid the tool point into cutting position. The Cut,4.25D will move the tool point, in feed, to 4.25 diameter, while the second cut statement will retract the tool point to clear the part.

Tool change and motion statements for thread cutting:

ATCHG,TOOL4,OFFSET4,GLX2.75,GLZ1,TLR,50FPM,RPMD4.5
THRD12,S(LN5/0.3ZL),F(LN7/0.1ZL),MID(LN6/0.06XS),MAD(LN6),DEG29.5, (cont)
SDPTH0.012,FDPTH0.01,2SP/0.001,BDLP2,CYC4/OFF,CO

THRDn initiates the threading cycle, and $n = 12$ tpi.
MIDn represents the thread minor diameter and n the material to remove.
MADn is the thread major diameter, and n indicates its value.
DEGn specifies the angle n of infeed for the tool in the threading statement.

SDPTHn is the starting depth of the first cut, with $n = 0.012$.
FDPTHn is the final depth of the last cut and $n = 0.01$.
BDLPn requires a block delete to be generated for the last $n = 2$ passes.
CYC4/OFF will assure that the tool will not return to home position before the last pass has been completed.
nSP/i represents $n = 2$ spring passes at $i = 0.001$ between the roughing and the last passes.

In order to better appreciate the extent of the complexity of this last statement, the reader should carefully study the tape file following tool change 4.

Tool change and motion statements for drilling:

ATCHG,TOOL5,OFFSET5,0.75TD,118TPA,GLX0,GLZ8,80FPM,0.006IPR
DRL,PT(XB,ZB),SDPTH0.75,FDPTH0.6,2.5DP,0.1CLEAR

SDPTHn specifies peck-drilling with $n = 0.75$ as first peck
FDPTHn specifies the subsequent pecking depth of $n = 0.6$.

END

It is strongly recommended that the reader study the source program in parallel with the list and the tape files to better understand the computer logic in processing the program.

The SOURCE PROGRAM is the NC computer program written by the programmer. This program consists of the start-up, geometry, tool change, and motion instructions. This program is punched on tape, read and stored on disk in an in-house Compact II computer. The program must then be processed and edited by the programmer off the terminal in an interactive conversational mode. The Compact II instruction word for processing is "Run" and the programmer answers questions from the computer by yes or no. In "run" mode, assisted by "substitute," the computer analyzes each statement for spelling, mathematical and motion accuracy, etc. If the run is successful, the computer will print out the total machining time and the length of tape in both feet and meters.

```
MACHIN,DEMO LATHE
IDENT, CNC-JP-11        FRONT LATHE
INIT,INCH/IN,INCH/OUT
SETUP,MOD2/1,CMOD2/2,ABS02,ZEROS3,RPM2500,X3+1+(3/64)+2.875
Z7.5+7,LIMIT(X0/10,Z2/18),FRONT
BASE,XA,7.5ZA
DLN1,ZB,2D,45CCW
DLN2,4.5D,-2.011-.25ZB,45CCW
DPT1,4.75D,-2-(11/32)-2.011ZB
```

```
DPT2,6D,-1.75-2-(11/32)-2.011ZB
DLN3,PT1,PT2
DLN4,2.5D
DLN5,-2.011ZB
DLN6,4.5D
DLN7,-2.011-2-(11/32)ZB
DLN8,6D
DLN9,ZB
DCIR1,LN5/.5ZL,LN4/.5XL,.5R
ATCHG,TOOL1,OFFSET1,GLX2.875,GLZ1.25,TLR(3/64),.008IPR,400FPM,RANGE1
MOVEC,OFFLN1/.35XL,OFFLN9/ZL
CUT,ONLN(XB),CSS/ON
CUT,.05X,.05Z
ATCHG,TOOL2,OFFSET2,GLX2.875,GLZ1.25,TLR(1/32),400FPM,.015IPR
MOVEC,OFFLN1/XL,OFFLN9/.1ZL
CUT,PARLN1,OFFLN4/XL,CSS/ON
ICON,CIR1,S(TANLN4),F(TANLN5),CCW
CUT,PARLN5,OFFLN2/XL
CUT,PARLN2,OFFLN6/XL
CUT,PARLN6,OFFLN7/ZL
CUT,PARLN7,OFFLN3/XL
CUT,PARLN3,OFFLN8/XL
MOVE,OFFLN9/.1ZL
ATCHG,TOOL3,OFFSET3,GLX2.5,GLZ1,TLR,CSS/OFF,80FPM,RPMD4.25,.004IPR
MOVE,OFFLN3/.05XL,OFFLN7/ZL
CUT,4.25D
CUT,OFFLN3/.05XL,PARLN7
ATCHG,TOOL4,OFFSET4,GLX2.75,GLZ1,TLR,50FPM,RPMD4.5
THRD12,S(LN5/.3ZL),F(LN7/.1ZL),MID(LN6/.06XS),MAD(LN6),DEG29.5,SDPTH.012
FDPTH.01,2SP/.001,BDLP2,CYC4/OFF,CO .
ATCHG,TOOL5,OFFSET5,.75TD,118TPA,GLZ8,GLX0,80FPM,.006IPR
DRL,PT(XB,ZB),SDPTH.75,FDPTH.6,2.5DP,.1CLEAR
END
```

The LIST FILE is compiled and assembled by the Compact II computer during the run mode. The list is printed at the request of the programmer. It will print each line, or group of lines, from the source program with the corresponding tape blocks of the machine control tape; its major reason for existing is to facilitate error correction by the programmer.

```
LIST
LIST FILE: /CNC-JP11L/ OK? YES


>MACHIN,DEMO LATHE
MAIN 60980 LINK 11079 SYS 100279 L# 1537

>IDENT, CNC-JP-11       FRONT LATHE
00-00-00  00:00

>INIT,INCH/IN,INCH/OUT

>SETUP,MOD2/1,CMOD2/2,ABS02,ZEROS3,RPM2500,X3+1+(3/64)+2.875
Z7.5+7,LIMIT(XO/10,Z2/19),FRONT

>BASE,XA,7.5ZA
= X . Z 7.5
```

```
>DLN1,ZB,2D,45CCW
= X 1. Z 7.5 A 45.

>DLN2,4.5D,-2.011-.25ZB,45CCW
= X 2.25 Z 5.239 A 45.

>DPT1,4.75D,-2-(11/32)-2.011ZB
= X 2.375 Z 3.1453

>DPT2,6D,-1.75-2-(11/32)-2.011ZB
= X 3. Z 1.3953

>DLN3,PT1,PT2
= X 2.375 Z 3.1453 A 19.6538

>DLN4,2.5D
= X 1.25 Z . A .

>DLN5,-2.011ZB
= X . Z 5.489 A 90.

>DLN6,4.5D
= X 2.25 Z . A .

>DLN7,-2.011-2-(11/32)ZB
= X . Z 3.1453 A 90.

>DLN8,6D
= X 3. Z . A .

>DLN9,ZB
= X . Z 7.5 A 90.

>DCIR1,LN5/.5ZL,LN4/.5XL,.5R
= X 1.75 Z 5.989 R .5

>ATCHG,TOOL1,OFFSET1,GLX2.375,GLZ1.25,TLR(3/64),.003IPR,400FPM,RANGE
 N0001 G90
 N0002 G94
 N0003 G04 X2. T0101
 N0004 G50 X 8.0938 Z 13.25

>MOVEC,OFFLN1/.35XL,OFFLN9/ZL
 N0005 G97 S0520 M40
 N0006 G01 X 3.0288 Z 7.5469 F300. M03

>CUT,ONLN(XB),CSS/ON
 N0007 G50 S0625
 N0008 G96 R 1.4676 S0400
 N0009 G95
 N0010 G01 X . F.008

>CUT,.05X,.05Z
 N0011 X .1 Z 7.5969

>ATCHG,TOOL2,OFFSET2,GLX2.875,GLZ1.25,TLR(1/32),400FPM,.015IPR
 N0012 G97 S0625
 N0013 G94
 N0014 G01 X 8.0938 Z 13.25 F300.
 N0015 G90
 N0016 G94
 N0017 G04 X2. T0202
 N0018 G50 X 8.0938 Z 13.25
```

```
>MOVEC,OFFLN1/XL,OFFLN9/.1ZL
  N0019 G97 S0625 M40
  N0020 G01 X 1.825 Z 7.6313 F300. M03

>CUT,PARLN1,OFFLN4/XL,CSS/ON
  N0021 G50 S0625
  N0022 G96 R .8818 S0400
  N0023 G95
  N0024 G01 X 2.5626 Z 7.2631 F.015

>ICON,CIR1,S(TANLN4),F(TANLN5),CCW
  N0025 Z 5.989
  N0026 G02 X 3.5 Z 5.52Q3 I .4687 K

>CUT,PARLN5,OFFLN2/XL
  N0027 G01 X 4.0258

>CUT,PARLN2,OFFLN6/XL
  N0028 X 4.5626 Z 5.252

>CUT,PARLN6,OFFLN7/ZL
  N0029 Z 3.1765

>CUT,PARLN7,OFFLN3/XL
  N0030 X 4.794

>CUT,PARLN3,OFFLN8/XL
  N0031 X 6.0626 Z 1.4006

>MOVE,OFFLN9/.1ZL
  N0032 G00 Z 7.6313

>ATCHG,TOOL3,OFFSET3,GLX2.5,GLZ1,TLR,CSS/OFF,80FPM,RPMD4.25,.004IPR
  N0033 G97 S0254
  N0034 G94
  N0035 G01 X 8.0938 Z 13.25 F300.
  N0036 G90
  N0037 G94
  N0038 G04 X2. T0303
  N0039 G50 X 8.8438 Z 13.5

>MOVE,OFFLN3/.05XL,OFFLN7/ZL
  N0040 G97 S0071 M40
  N0041 G01 X 4.8562 Z 3.1453 F300. M03

>CUT,4.25D
  N0042 G95
  N0043 G01 X 4.25 F.004

>CUT,OFFLN3/.05XL,PARLN7
  N0044 X 4.8562

>ATCHG,TOOL4,OFFSET4,GLX2.75,GLZ1,TLR,50FPM,RPMD4.5
  N0045 G94
  N0046 G01 X 8.8438 Z 13.5 F300.
  N0047 G90
  N0048 G94
  N0049 G04 X2. T0404
  N0050 G50 X 8.3438 Z 13.5

>THRD12,S(LN5/.3ZL),F(LN7/.1ZL),MID(LN6/.06XS),MAD(LN6),DEG29.5,SDPTH.02
 FDPTH.01,ZSP/.001,BDLP2,CYC4/OFF,CO *
```

```
N0051 G97 S0062 M40
N0052 G01 X 4.6 Z 5.789 F300. M03
N0053 X 4.4814 Z 5.7554 F.25
N0054 G33 Z 3.2453 K.08333
N0055 X 4.6 I.08333
N0056 G00 Z 5.789
N0057 G01 X 4.4624 Z 5.7501
N0058 G33 Z 3.2453 K.08333
N0059 X 4.6 I.08333
N0060 G00 Z 5.789
N0061 G01 X 4.4432 Z 5.7446
N0062 G33 Z 3.2453 K.08333
N0063 X 4.6 I.08333
N0064 G00 Z 5.789
N0065 G01 X 4.4238 Z 5.7391
N0066 G33 Z 3.2453 K.08333
N0067 X 4.6 I.08333
N0068 G00 Z 5.789
N0069 G01 X 4.404 Z 5.7336
N0070 G33 Z 3.2453 K.08333
N0071 X 4.6 I.08333
N0072 G00 Z 5.789
N0073 G01 X 4.384 Z 5.7279
N0074 G33 Z 3.2453 K.08333
N0075 X 4.6 I.08333
N0076 G00 Z 5.789
N0077 G01 X 4.382 Z 5.7273
N0078 G33 Z 3.2453 K.08333
N0079 X 4.6 I.08333
N0080 G00 Z 5.789
N0081 G01 X 4.38 Z 5.7268
N0082 G33 Z 3.2453 K.08333
N0083 X 4.6 I.08333

>ATCHG, TOOL5, OFFSET5, .75TD, 118TPA, GLZ8, CLXO, 80FFM, .006IPR
N0084 G90
N0085 G94
N0086 G01 X 8.3438 Z 13.5 F300.
N0087 G90
N0088 G94
N0089 G04 X2. T0505
N0090 G50 X 13.8438 Z 6.5

>DRL, PT(XB, ZB), SDPTH.75, FDPTH.6, 2.5DP, .1CLEAR
N0091 G97 S0407 M40
N0092 G00 X 13.8438 Z 7.6 M03
N0093 X .
N0094 G95
N0095 G01 Z 7.0098 F.006
N0096 G00 Z 7.6
N0097 Z 7.0598
N0098 G01 Z 6.4922
N0099 G00 Z 7.6
N0100 Z 6.5422
N0101 G01 Z 5.9472
N0102 G00 Z 7.6
N0103 Z 5.9972
N0104 G01 Z 5.3747
N0105 G00 Z 7.6
N0106 Z 5.4247
N0107 G01 Z 4.7747
N0108 G00 Z 7.6
```

```
> END
  NO109  G94
  NO110  G01  X 13.8438  Z 6.5 F300.
  NO111  G04  X1.  T0500
  NO112  M30
END MIN: 12.8  FT: 24.8  MTR: 7.5

+
```

The TAPE FILE is produced at the same time, provided there are no errors during processing. This file should be printed and checked prior to punching the CNC tape. Changes to this file can be made in "edit" mode, thus providing the knowledgeable programmer with an additional level of flexibility. Once the programmer is satisfied with the output, the computer can be instructed to punch out the machine control tape. The programmer has the option to punch the tape, punch and print the file or print only.

```
  TAPE
TAPE FILE: /CNC-JP11T/ OK? YES
EIA? NO
OUTPUT TO: TPT
TURN PUNCH ON,HIT CR

 +BBB<+JJBB+   +<B+B<+>HHH>@@+@@+++JJBB<BBB<<BBB<<BBB<<BBB<<BBB<<BBB
N0001G90
N0002G94
N0003G04X2.T0101
N0004G50XS.0938Z13.25
N0005G97S0520M40
N0006G01X3.0298Z7.5469F300.M03
N0007G50S0625
N0008G96R1.4676S0400
N0009G95
N0010G01X.F.008
N0011X.1Z7.5969
N0012G97S0625
N0013G94
N0014G01X8.0938Z13.25F300.
N0015G90
N0016G94
N0017G04X2.T0202
N0018G50X8.0938Z13.25
N0019G97S0625M40
N0020G01X1.826Z7.6313F300.M03
N0021G50S0625
N0022G96R.8818S0400
N0023G95
N0024G01X2.5626Z7.2631F.015
N0025Z5.989
N0026G02X3.5Z5.5203I.4687K
N0027G01X4.0258
N0028X4.5626Z5.252
N0029Z3.1765
N0030X4.794
N0031X6.0626Z1.4006
N0032G00Z7.6313
```

```
N0033G97S0254
N0034G94
N0035G01X8.0938Z13.25F300.
N0036G90
N0037G94
N0038G04X2.T0303
N0039G50X3.8438Z13.5
N0040G97S0071M40
N0041G01X4.8562Z3.1453F300.M03
N0042G95
N0043G01X4.25F.004
N0044X4.8562
N0045G94
N0046G01X8.8438Z13.5F300.
N0047G90
N0048G94
N0049G04X2.T0404
N0050G50X8.3438Z13.5
N0051G97S0062M40
N0052G01X4.6Z5.789F300.M03
N0053X4.4814Z5.7554F.25
N0054G33Z3.2453K.08333
N0055X4.6I.08333
N0056G00Z5.789
N0057G01X4.4624Z5.7501
N0058G33Z3.2453K.08333
N0059X4.6I.08333
N0060G00Z5.789
N0061G01X4.4432Z5.7446
N0062G33Z3.2453K.08333
N0063X4.6I.08333
N0064G00Z5.789
N0065G01X4.4238Z5.7391
N0066G33Z3.2453K.08333
N0067X4.6I.08333
N0068G00Z5.789
N0069G01X4.404Z5.7336
N0070G33Z3.2453K.08333
N0071X4.6I.08333
N0072G00Z5.789
N0073G01X4.384Z5.7279
N0074G33Z3.2453K.08333
N0075X4.6I.08333
N0076G00Z5.789
N0077G01X4.382Z5.7273
N0078G33Z3.2453K.08333
N0079X4.6I.08333
N0080G00Z5.789
N0081G01X4.38Z5.7268
N0082G33Z3.2453K.08333
N0083X4.6I.08333
N0084G90
N0085G94
N0086G01X8.3438Z13.5F300.
N0087G90
N0088G94
N0089G04X2.T0505
N0090G50X13.8438Z6.5
N0091G97S0407M40
N0092G00X13.8438Z7.6M03
N0093X.
N0094G95
```

```
N0095G01Z7.0098F.006
N0096G00Z7.6
N0097Z7.0598
N0098G01Z6.4922
N0099G00Z7.6
N0100Z6.5422
N0101G01Z5.9472
N0102G00Z7.6
N0103Z5.9972
N0104G01Z5.3747
N0105G00Z7.6
N0106Z5.4247
N0107G01Z4.7747
N0108G00Z7.6
N0109G94
N0110G01X13.8438Z6.5F300.
N0111G04X1.T0500
N0112M30
```

12.9   COMPUTER-AIDED MANUFACTURING—CAM

Due to recent developments in computer technology, the cost of computation has decreased approximately tenfold in the last decade. This reduced cost factor has initiated an unprecedented rate of change in computer-aided design and computer-aided manufacturing. During the 1980s, productivity will be the most important factor to be faced by manufacturers, with competition from external sources. A new wave of worldwide industrial automation is in progress, based on rapidly increasing use of computers in design and manufacturing. This presents both a challenge and an opportunity to utilize adaptive controls and robotics as part of a computer-integrated system.

Appendix

TABLE A-1
MISCELLANEOUS FUNCTIONS (M-CODES)

M-Code	Name
M 00	Program stop
M 01	Optional stop
M 02	End of program and tape rewind
M 03	Spindle start CW
M 04	Spindle start CCW
M 05	Spindle stop
M 06	Tool change
M 08	Coolant ON
M 09	Coolant OFF
M 19	Spindle orient and stop
M 21	Mirror image X
M 22	Mirror image Y
M 23	Mirror image OFF
M 30	End of program and memory rewind
M 41	Low range
M 42	High range
M 48	Override cancel OFF
M 49	Override cancel ON
M 98	Go to subroutine
M 99	Return from subroutine

Note: Numerous other M-codes are available on various machines, as indicated by the respective manufacturer.

TABLE A-2 PREPARATORY FUNCTIONS (G-CODES) IN MILLING	
G-Code	**Name**
G00	Positioning (Rapid traverse)
G01	Linear interpolation (Cutting feed)
G02	Circular interpolation (Clockwise)
G03	Circular interpolation (Counterclockwise)
G04	Dwell cycle
G10	Offset value setting by program
G17	XY plane selection
G18	ZX plane selection
G19	YZ plane selection
G20	Inch data input (G70 on some systems)
G21	Metric data input (G71 on some systems)
G22	Safety zone programming
G23	Programmed crossing through safety zone
G27	Reference point return check
G28	Return to reference point
G29	Return from reference point
G30	Return to 2nd reference point
G40	Cutter diameter compensation cancel
G41	Cutter diameter compensation left
G42	Cutter diameter compensation right
G43	Tool length compensation + direction
G44	Tool length compensation − direction
G49	Tool length compensation cancel
G45	Tool offset increase

TABLE A-2 (*continued*)	
G-Code	**Name**
G46	Tool offset decrease
G47	Tool offset double increase
G48	Tool offset double decrease
G73	Peck-drilling cycle
G74	Counter tapping cycle
G76	Fine boring cycle
G80	Canned cycle cancel
G81	Drilling cycle, spot drilling cycle
G82	Drilling cycle, counter boring cycle
G83	Peck-drilling cycle
G84	Tapping cycle
G85	Boring cycle
G86	Boring cycle
G87	Back boring cycle
G88	Boring cycle
G89	Boring cycle
G90	Absolute input
G91	Incremental input
G92	Programming of absolute zero point
G98	Return to initial level
G99	Return to R level

Note: Always follow the operator's and programmer's manual for *your* system.

TABLE A-3 PREPARATORY FUNCTIONS (G-CODES) IN TURNING		
Standard G code	Special G code	Name
G00	G00	Positioning (Rapid traverse)
G01	G01	Linear interpolation (Cutting feed)
G02	G02	Circular interpolation (Clockwise)
G03	G03	Circular interpolation (Counterclockwise)
G04	G04	Dwell cycle
G10	G10	Offset value setting by program
G20	G20	Inch data input (G70 on some systems)
G21	G21	Metric data input (G71 on some systems)
G22	G22	Safety zone programming
G23	G23	Programmed crossing through safety zone
G27	G27	Reference point return check
G28	G28	Return to reference point
G29	G29	Return from reference point
G30	G30	Return to 2nd reference point
G31	G31	Skip cutting
G32	G32	Thread cutting
G34	G34	Variable lead thread cutting
G36	G36	Automatic tool compensation
G37	G37	Automatic tool compensation
G40	G40	Tool nose radius compensation cancel
G41	G41	Tool nose radius compensation left
G42	G42	Tool nose radius compensation right
G50	G92	Programming of absolute zero point Maximum spindle speed setting

Standard G-Code	Special G code	Name
		TABLE A-3 (*continued*)
G65	G65	User macro call command
G66	G66	Modal user macro call command
G67	G67	Modal user macro call cancellation
G68	G68	Mirror image for double turrets ON
G69	G69	Mirror image for double turrets OFF
G70	G70	Finishing cycle
G71	G71	Stock removal in turning
G72	G72	Stock removal in facing
G73	G73	Pattern repeating
G74	G74	Peck-drilling in Z axis
G75	G75	Grooving in X axis
G76	G76	Thread cutting cycle
G90	G77	Cutting cycle A
G92	G78	Thread cutting cycle
G94	G79	Cutting cycle B
G96	G96	Constant surface speed control on
G97	G97	Constant surface speed control cancel
G98	G94	Feed per minute
G99	G95	Feed per revolution
—	G90	Absolute programming
—	G91	Incremental programming

Note: Always follow the operator's and programmer's manual for *your* system.

Index